Introduction to Medical Electronics Applications

Introduction to Medical Electronics Applications

D. Jennings, A. Flint, B.C.H. Turton and L.D.M. Nokes

School of Engineering
University of Wales, College of Cardiff

Edward Arnold
A member of the Hodder Headline Group
LONDON   BOSTON   SYDNEY   AUCKLAND

First published in Great Britain in 1995 by
Edward Arnold, a division of Hodder Headline PLC,
338 Euston Road, London NW1 3BH

Distributed in the USA by
Little, Brown and Company
34 Beacon Street, Boston, MA 02108

Whilst the advice and information in this book is believed to be true and
accurate at the date of going to press, neither the author nor the publisher
can accept any legal responsibility or liability for any errors or omissions
that may be made. In particular (but without limiting the generality of the
preceding disclaimer) every effort has been made to check drug dosages;
however, it is still possible that errors have been missed. Furthermore,
dosage schedules are constantly being revised and new side effects
recognised. For these reasons the reader is strongly urged to consult the
drug companies' printed instructions before administering any of the drugs
recommended in this book.

British Library Cataloguing in Publication Data
A catalogue record for this book is available from the British Library

ISBN 0 340 61457 9

1 2 3 4 5   95 96 97 98 99

Typeset in Times by GreenGate Publishing Services, Tonbridge, Kent
Printed and bound in Great Britain by J.W. Arrowsmith Ltd., Bristol

Contents

Preface

This book is intended as an introductory text for Engineering and Applied Science Students to the Medical Applications of Electronics. A course has been offered for many years in Cardiff in this arena both in this College and its predecessor institution. A new group, the Medical Systems Engineering Research Unit, was established following the reorganisation of the College. Restructuring and review of our course material and placing the responsibility for teaching this course within the new group led to a search for new material. Whilst we found a number of available texts which were suitable for aspects of our new course, we found a need for a text which would encompass a wide scope of material which would be of benefit to students completing their degree programmes and contemplating professional involvement in Medical Electronics.

Medical Electronics is a broad field. Whilst much of the material which an entrant to medical applications must acquire is the conventional basis of electronics covered by any student of electronics, there are areas of special emphasis. Many of these arise from areas which are increasingly inaccessible to students who necessarily specialise at an early stage in their education.

The need for diversity is reflected in the educational background and experience of the authors. Amongst us is a Medical Practitioner who is also a Mechanical Engineer, a Physicist who now works as a Software Engineer, an Electronics Engineer who made the same move, and another Electronics Engineer with some experimental experience in Orthopaedics.

The material which this book attempts to cover starts with an Introduction which hopefully provides some perspective in the subject area. The following chapter provides an introduction to human anatomy and physiology. The approach taken here is necessarily simplified: it is our intention to provide an adequate grounding for the material in the following chapters both in its basic science and the nomenclature which may be unfamiliar to readers with only elementary biological knowledge.

Chapter Three describes the Physics employed in diagnostic techniques. This encompasses basic radiation physics, magnetic resonance and the nature and generation of ultrasound. Chapter 4 discusses the form of some of the basic electronic elements used in Medical Applications. We describe the specialised techniques which are employed and characterise the signals which are likely to be encountered. Special emphasis is attached to issues of patient safety, although these are covered in greater depth in Chapter 8.

The mathematical background for image processing is covered in Chapter 5. This material has been separated from our description of representative diagnostic imaging technologies presented in Chapter 6. This latter Chapter includes material supplied by Toshiba Medical Systems, whose assistance we gratefully acknowledge.

Chapter 7 contains background material concerning computers, their architecture, application to data acquisition and connection to networks. It also covers some aspects of the application of Databases and Expert Systems to Medicine which have long been expected to play central roles in patient care. The increasing capacity of systems together with their continuing cost reductions mean that their introduction is now becoming a reality. The introductory parts of this Chapter will be familiar to many engineers: we have included it to ensure that this book shall have a wide enough sphere of interest.

Finally, Chapter 8 examines aspects of patient safety which are of concern to engineers. This area is a particularly difficult one in which to be specific as it is intimately entwined with changing legislation. We seek to present here principles and what we believe to be good practice: these must form the basis of any competent engineer's activity.

This book has been some time in gestation. We wish to acknowledge the patience of our families, without whom no doubt the task would have been completed more quickly. We have been assisted too in no small measure by students and researchers in the Medical Systems Engineering Research Unit who have provided both constructive criticisms and help by checking manuscripts.

1

Introduction

This book is concerned with describing the application of technological methods to medical diagnosis and therapy. It is instructive to review its development through recorded history. It is apparent that the fastest advances in the application of technology to medicine have occurred in the 20th Century and with an increasing pace. The following paragraphs touch on some events in this chain. We should recall that systematic technological assistance has only recently been widely applied to medicine through engineering. An understanding of the pathology which technology often helps to identify has largely been developed hand in hand with its application. In these paragraphs, we identify a number of the technologically based systems which are described more fully in the succeeding chapters: their descriptions here are necessarily rather terse.

Medicine arose as a *Scientific* discipline in ancient times. Bernal (1957) notes that by the time of the establishment of the Greek civilisation, physicians were a notable professional group whose activities were essential to the affluent, partly as a result of their unhealthy lifestyle. They had by the 3rd Century BC distinguished between sensory and motor nervous functions. In the same era the Hippocratic Oath, or the code of conduct for physicians was written: it remains today as an ethical basis for much of medical practice.

Spectacles are first described in mid 14th Century Italy. Whilst optical glass had been used for a long period, the quality of glass used by the ancients was too flawed to be of use for eyesight correction. The continuing development of spectacle lenses led by about 1600 to the development of the first telescopes. By the Renaissance period in the early 15th Century, medicine was becoming more formalised. Anatomical knowledge progressively improved, and although the topics of pathology and physiology were recognised, they had advanced little from the time of Galen in Second Century Greece. Modern scientific medicine based on biological science has largely developed since the mid 19th Century work by Pasteur and others. Bernal (1957) notes that they provided the theories which led to an understanding of epidemiology and to rational descriptions of nervous function.

The practical development of a thermometer suitable for measurement of body temperature dates back to 1625. Whilst internal sounds from the body have been observed by physicians since the time of the Romans, the stethoscope dates back to the 19th Century, in a form reasonably similar to the present.

Whilst crafted artificial replacements for severed limbs have been in use for many centuries, the development of both implanted prosthesis and functional artificial limbs is recent.

The measurement of the electrical signals carried by our nervous system (known as Biopotentials) dates from the early years of the 20th Century with the first measurements of the Electrocardiograph. By the 1940s paper chart recordings of the detected waveforms could be made. The same era saw the development of the use of Electrosurgery, which employs resistive heating either to make delicate incisions or to cauterise a wound. By the 1960s, electrical stimulation of the heart was employed, firstly in the defibrillator either to restart or resynchronise a failing heart, and secondly in miniaturised pacemakers which could be used in the long term to bypass physical damage to parts of the heart. Electricity has also been applied, perhaps more controversially, since the 1940s in Electroconvulsive Therapy (ECT) to attempt to mitigate the effects of a number of psychiatric conditions.

Apart from sensing signals generated by the body, clinical medicine has been greatly advanced by the use of imaging techniques. These afford the possibility of viewing structures of the body which are otherwise inaccessible. They may either operate on a scale which is characterised by the transfer of chemicals or on a structural level, perhaps to examine the fracture of a bone.

X rays have been applied to diagnosis since soon after their discovery by Röntgen in 1895. The source of diagnostic radiation was the Cathode Ray Tube (CRT) which produced penetrating photons which could be viewed on a photographic emulsion. The early days of the 20th Century saw the first use of ionising radiation in Radiotherapy for the treatment of cancerous conditions. A failure to appreciate the full extent of its dangers led to the premature deaths of many of its early proponents. Early medical images were recorded using the ancestors of the familiar X ray films. However, since the 1970s, acquisition of radiographic data using electronic means has become progressively more commonplace. The newer technique affords the possibility of processing the image to 'improve' aspects of it, or enable its registration with other images taken at another time to view the progress of a condition.

A major technique for the visualisation of anatomical structures and the metabolism has been the use of radionuclides introduced into the body. The technology, known as Nuclear Medicine, has been used since about 1948 when radioactive iodine was first used to help examine the thyroid. The resolution available from nuclear medicine has progressively increased with increasing miniaturisation of the photomultiplier tubes used in its detectors and improvements to collimators.

Computerised Tomography has developed from its initial application as a medical diagnostic technique in 1972. It had an earlier history when many aspects of the technique were demonstrated although without medical application. The use of computerised tomography has been one of the signal events in the development of medical imaging, enabling views of internal structures of a quality hitherto impossible. The technique has been refined somewhat from its inception in terms of degree: the time to obtain an image has significantly been accelerated and thereby provided commensurate reductions in patient radiation dose. Processing of the images obtained has also moved forward dramatically enabling three dimensional images to be obtained and presented with an illusion of perspective.

Much of the work in image processing in general owes its origins to fields outside of medicine. The mathematics developed for image analysis of astronomical data has been applied to contribute to a number of aspects of medical image processing. In order to be of reasonably general use, images should ideally provide representations of the systems which they examine in terms which are accessible to a non-specialist. The early projection X ray images are characterised by information accumulated from the summation of absorption of radiation along the paths of all rays. The resulting image does not represent the morphology of a single plane or structure but instead is a complex picture of all the contributing layers. This requires a high degree of skill to interpret. Image processing may help in ways such as clarifying the data of interest, removing movement artefacts and providing machine recognition of certain structures. These functions enable the extension of the application of medical imaging to the quantification of problems such as the stroke volume of the heart so that its operation may be properly assessed whilst minimising the use of invasive techniques.

Another technique which has been applied to medicine in the recent past and with increasing success is ultrasonic diagnosis. This arose from two fields. The first was the application of sonar in the Second World War to submarine location. Also developed during the War was Radar: this relies on very a similar mathematical basis to obtain images by what is essentially the reflection of a portion of the energy from a source back to a detector. The development of signal processing for radar has been one of the major early inputs into the development of medical ultrasonic diagnosis systems. A significant difference in difficulty of analysis of their respective signals is due to the much greater non-uniformity of the medium through which ultrasound is passed. Ultrasound diagnostic systems are now in widespread use, particularly in applications such as gynaecology in which the hazards due to ionising radiation present an unacceptable risk for their routine use. Gynaecological screening by ultrasound is undertaken now routinely in many countries: although doubts about its absolute safety have been expressed, no causative links to ailments have yet been established.

Ultrasound also provides a suitable mechanism for use with Doppler techniques, again borrowed substantially from radar, to measure the velocities of blood or structures. Doppler ultrasonic examinations provide a safe non-invasive means for the measurement of cardiovascular function which previously required the use of much more hazardous techniques including catheterisation.

Since the early 1980s there has been a rapid introduction of the medical application of Nuclear Magnetic Resonance (NMR). The physical phenomenon was first described in 1946, and was able to determine the concentrations of certain chemicals in samples. In the application in medicine it is able to provide three dimensional discrimination of the positions of concentrations of the nuclei of atoms which have characteristic spins: in particular the location of hydrogen nuclei may be recognised. The information obtained by NMR is called Magnetic Resonance Imaging, or MRI, in its medical application. The images provide an excellent resolution and discrimination between many corporeal structures. They are obtained without known deleterious effects in most cases, although the equipment required to obtain MRI images costs significantly more than that required for other image acquisition mechanisms, known as *modalities*.

The development of electronics, and particularly that of computers has made possible many of the technologies which we shall examine.

Firstly, computers are the central elements involved in processing signals in many cases, and particularly those obtained from images. The special nature of the processing required to obtain the image improvements required and the consequential flexibility in their application mean that the complexity of the algorithms for processing would be excessive unless software was used for managing the process. Medical image processing frequently requires that different views may need to be synthesised in the examination of a condition relating to each particular patient. The exact form of the views may be difficult to predict, so computers provide the ideal platform for their analysis.

Secondly the increasing use of computers in medical applications has led to an ever increasing capability to retain medical data. This may be used to facilitate health care planning and to provide for a reliable storage of patient related data which may be readily recovered. They also provide the ability to communicate data using standardised mechanisms which we may expect will increasingly allow data to be acquired in one location and viewed at another.

Finally computers have potential for providing us with systems which mimic the diagnostic processes employed by physicians. Pilot systems which can provide some diagnostic assistance have been tried for a number of years in certain areas both within and outside medicine. They are particularly prevalent in manufacturing industry where they may be employed to assist in the design process and to control the flow of goods through factories. Clearly such systems are limited in their scope by the complexity of their programming. We should also not forget that humans undertake certain tasks particularly well, such as pattern recognition of faces as a result of possibly innate training.

We should end this overview of the application of technology to medicine by considering two things.

1. When we contemplate applying a technological solution to a problem, will it benefit the patient? The benefit may either be direct in terms of an immediate improvement in the patient's condition, or one which facilitates action as a result of time saving. A computer may, in some circumstances, undertake a task either much more quickly, or more reliably than a human. On the other hand, there are many cases when the computer's instructions have not been formulated in a manner which enable it to handle the task at all.

2. Will the application provide a global benefit, or is it likely to result in some other detrimental effect? In cases where technology is used without considering all its effects, it frequently transpires that the task could have been undertaken more simply. Much more seriously, the problem may be reflected by placing excessive reliance on a technological solution in an inappropriate manner. We must be particularly confident when we hand a safety critical task to a machine that we retain a sufficient view and knowledge of the problem in order to take appropriate action should unforeseen circumstances arise. In other words we should not always be excessively comforted by the reliability of the apparatus to lull us into a false sense of security.

2

Anatomy and Physiology

2.1. Introduction

Before proceeding to the various anatomical levels that can be found in the human body, it would be useful to have some simple definitions. The definition of anatomy is the study of structures that make up the human body and how they relate to each other, for example, how does the skeletal structure relate to the muscular structure, or how does the cardiovascular structure relate to the respiratory structure?

The definition for physiology is the study of the function of body structures, for example, how do the neural impulses transmit down a nerve and affect the structuring at the end of the nerve. In understanding these interactions, the application of electronics to monitor these systems will be more readily understood.

To describe the location of particular parts of the body, anatomists have defined the anatomical position. This is shown in Figure 2.1.

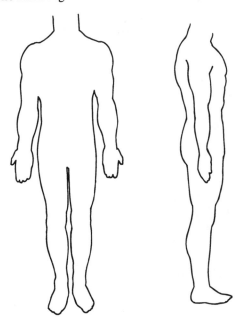

Figure 2.1 Anatomical position

2.2. Anatomical Terminology

There is standardised terminology to describe positions of various parts of the body from the midline. These are shown in Figure 2.2. When the body is in the 'anatomical position', it can be further described with relation to body regions. The main regions of the body are the axial, consisting of the head and neck, chest, abdomen and pelvis; the appendicular, which includes the upper extremities – shoulders, upper arms, forearms, wrists and hands; and the lower

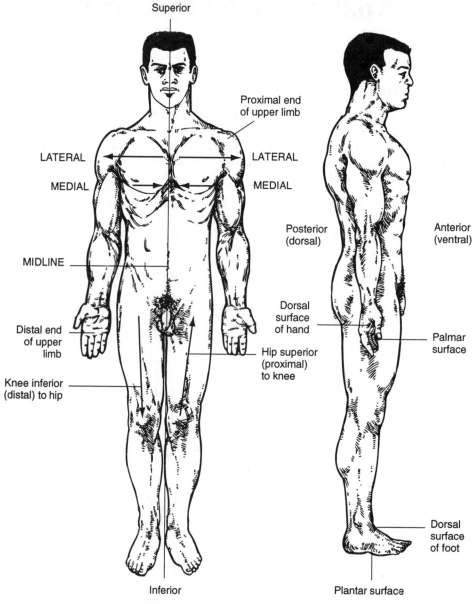

Figure 2.2 Standard body positions

extremities – hips, thighs, lower legs, ankles and feet. These are shown in Figure 2.3. Further subdivision in order to identify specific areas of the body can be carried out by considering various planes. These are shown in Figure 2.4. The midsagital plane divides the left and right sides of the body lengthwise along the midline. If the symmetrical plane is placed off centre and separates the body into asymmetrical left and right sections it is called the sagital plane. If you face the side of the body and make a lengthwise cut at right angles to the midsagital plane you would make a frontal (coronal) plane, which divides the body into asymmetrical anterior and posterior sections. A transverse plane divides the body horizontally into upper (superior) and lower (inferior) sections. An understanding of these terminologies is important, as it is the common language for locating parts in the human body. Without these definitions, confusion would arise in describing the relationship between one body part and another.

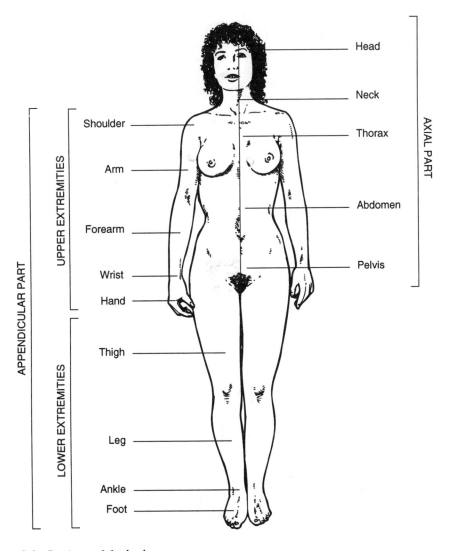

Figure 2.3a Regions of the body

2.3. Structural Level of the Human Body

The cell is assumed to be the basic living unit of structure of all organisms. Also, all living things are made up of one or more cells. Life is thought not to exist before the formation of a cellular structure.

Figure 2.5 is an example of a human cell. Although a very complex structure, it can be broken down into a number of components that interact with each other in order to perform specific functions required for life. In the centre of the cell is the nucleus. This is considered to be the control area that interacts with various parts of the cell body in order to maintain the cell's existence. The nucleus is bathed in a fluid called the cytoplasm. This is the factory of the cell and it is where components are manufactured on the instruction of the nucleus via chemical messengers, again to maintain the cellular function and existence.

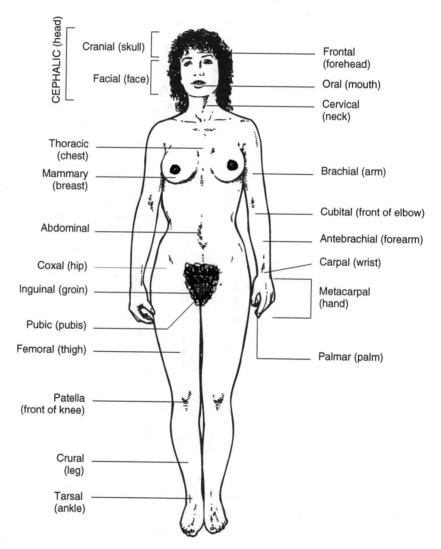

Figure 2.3b Regions of the body

The cell has to communicate with its environment. This is done via the plasma membrane, which lines the whole cell. Messengers in the form of molecules can be transmitted across this membrane, as it is permeable to specific molecules of various shapes and sizes. Movement of these messengers across the membrane is achieved by two mechanisms.

1. Simple diffusion: molecules pass through the membrane from high to low concentrations.

2. Active diffusion: basic fuel for the human body is adenosine triphosphate (ATP). This fuel acts on a pump that pushes molecules from a low concentration to a high concentration.

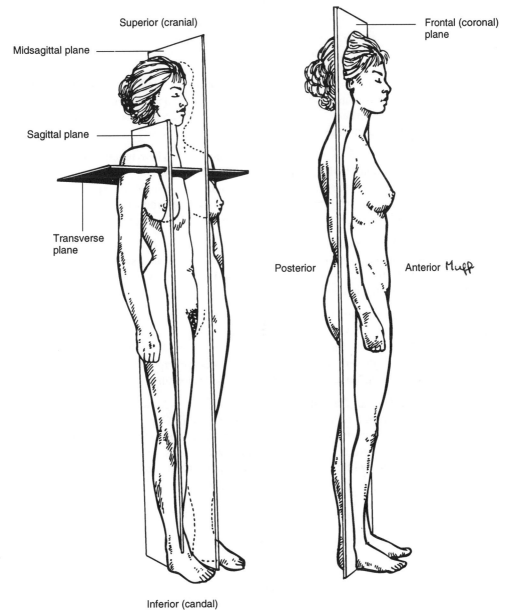

Figure 2.4 Body planes

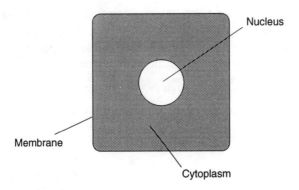

Figure 2.5 Schematic of human cell

When many similar cells combine to perform a specific function, they are called tissues. Examples of human tissue are epithelial, connective, muscle and nervous. It is important to stress that the difference between tissues is that the cells combine to perform a specific function associated with each tissue.

Epithelial tissues line all body surfaces, cavities and tubes. Their function is to act as an interface between various body compartments. They are involved with a wide range of activities, such as absorption, secretion and protection. For example, the epithelial lining of the small intestine is primarily involved in the absorption of products of digestion, but the epithelium also protects it from noxious intestinal contents by secreting a surface coating.

Connective tissue is the term applied to the basic type of tissue which provides structural support for other tissue. Connective tissue can be thought of as a spider's web that holds together other body tissues. Within this connective tissue web, various cells that fight the bacteria which invade the body can be found. Similarly, fat is also stored in connective tissue.

An organ is an amalgamation of two or more kinds of tissue that work together to perform a specific function. An example is found in the stomach; epithelial tissue lines its cavity and helps to protect it. Smooth muscle churns up food, breaks it down into smaller pieces and mixes it with digestive juices. Nervous tissue transmits nerve impulses that initiate the muscle contractions, whilst connective tissue holds all the tissues together.

The next structural level of the body is called systems. The system is a group of organs that work together to perform a certain function. All body systems work together in order that the whole body is in harmony with itself. Listed in Table 2.1 are the body systems and their major functions. Systems that are often monitored in order to analyse the well-being of the body include those associated with respiratory, skeletal, nervous and cardiovascular.

Table 2.1 Body Systems

The structures of each system are closely related to their functions.

Body system	Major functions
CARDIOVASCULAR (heart, blood, blood vessels)	Heart pumps blood through vessels; blood carries materials to tissues; transports tissue wastes for excretion.
DIGESTIVE (stomach, intestines, other digestive structures)	Breaks down large molecules into small molecules that can be absorbed into blood, removes solid wastes.
ENDOCRINE (ductless glands)	Endocrine glands secrete hormones, which regulate many chemical actions within the body.
INTEGUMENTARY (skin, hair, nails, sweat	Covers and protects internal organs; helps regulate body temperature.
LYMPHATIC (glands, lymph nodes, lymph, lymphatic vessels) and oil glands)	Returns excess fluid to blood; part of immune system.
MUSCULAR (skeletal, smooth cardiac muscle)	Allows for body movement; produces body heat.
NERVOUS (brain, spinal cord; peripheral nerves; sensory organs)	Regulates most bodily activities; receives and interprets information from sensory organs; initiates actions by muscles.
REPRODUCTIVE (ovaries, testes, reproductive cells, accessory glands, ducts)	Reproduction.
RESPIRATORY (airways, lungs)	Provides mechanism for breathing, exchange of gases between air and blood.
SKELETAL (bones, cartilage)	Supports body, protects organs; provides lever mechanism for movement; produces red blood cells.
URINARY (kidneys, ureters, bladder, urethra)	Eliminates metabolic wastes; helps regulate blod pressure, acid-base and water-salt balance.

Derived from Carola *et al.*, 1990

2.4. Muscular System

The function of muscle is to allow movement and to produce body heat. In order to achieve this, muscle tissue must be able to contract and stretch. Contraction occurs via a stimulus from the nervous system. There are three types of muscle tissue; smooth, cardiac and skeletal.

Skeletal muscle by definition is muscle which is involved in the movement of the skeleton. It is also called striated muscle as the fibres, which are made up of many cells, are composed of alternating light and dark stripes, or striations. Skeletal muscle can be contracted without conscious control, for example in sudden involuntary movement.

Most muscle is in a partially contracted state (tonus). This enables some parts of the body to be kept in a semi-rigid position, i.e. to keep the head erect and to aid the return of blood to the heart. Skeletal muscle is composed of cells that have specialised functions. They are called muscle fibres, due to their appearance as a long cylindrical shape plus numerous nuclei. Their lengths range from 0.1 cm to 30 cm with a diameter from 0.01 cm to 0.001 cm. Within these

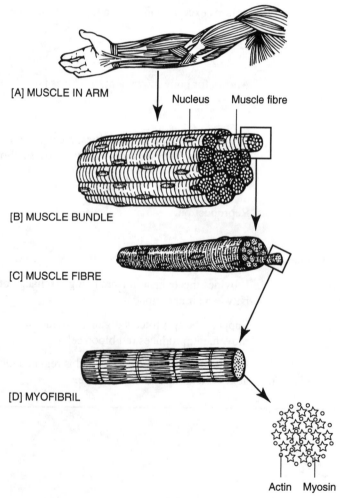

[A] MUSCLE IN ARM

Nucleus   Muscle fibre

[B] MUSCLE BUNDLE

[C] MUSCLE FIBRE

[D] MYOFIBRIL

Actin   Myosin

Figure 2.6 Gross to molecular structure of muscle

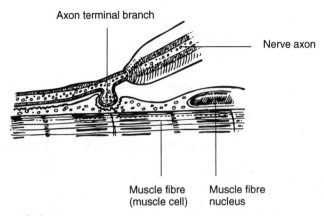

Axon terminal branch

Nerve axon

Muscle fibre
(muscle cell)

Muscle fibre
nucleus

Figure 2.7 Motor end plate

muscle fibres are even smaller fibres called myofibrils. These myofibrils are made up of thick and thin threads called myofilaments. The thick myofilaments are called myocin and the thin myofilaments are called actin. Figure 2.6 shows a progression from the gross to the molecular structure of muscle.

Control of muscle is achieved via the nervous system. Nerves are attached to muscle via a junction called the motor end plate. Shown in Figure 2.7 is a diagrammatic representation of a motor end plate.

2.4.1. Mechanism of Contraction of Muscle

Muscle has an all or none phenomenon. In order for it to contract it has to receive a stimulus of a certain threshold. Below this threshold muscle will not contract; above this threshold muscle will contract but the intensity of contraction will not be greater than that produced by the threshold stimulus.

The mechanism of contraction can be explained with reference to Figure 2.8. A nerve impulse travels down the nerve to the motor end plate. Calcium diffuses into the end of the nerve. This releases a neuro transmitter called acetylcholine, a neural transmitter. Acetylcholine travels

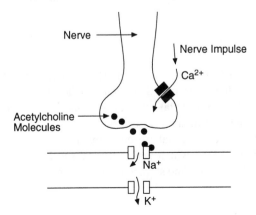

Nerve

Nerve Impulse

Ca^{2+}

Acetylcholine
Molecules

Na^+

K^+

Figure 2.8 Mechanism of muscle contraction

across the small gap between the end of the nerve and the muscle membrane. Once the acetylcholine reaches the membrane, the permeability of the muscle to sodium (Na^+) and potassium (K^+) ions increases. Both ions are positively charged. However, there is a difference between permeabilities for the two ions. Na^+ enters the fibre at a faster rate than the K^+ ions leave the fibre. This results in a positive charge inside the fibre. This change in charge initiates the contraction of the muscle fibre.

The mechanism of contraction involves the actin and myocin filaments which, in a relaxed muscle, are held together by small cross bridges. The introduction of calcium breaks these cross bridges and allows the actin to move using ATP as a fuel. Relaxation of muscle occurs via the opposite mechanism. The calcium breaks free from the actin and myocin and enables the cross bridges to reform. Recently there has been a new theory of muscle contraction. This suggests that the myocin filaments rotate and interact with the actin filaments, similar to a corkscrew action, with contacts via the cross bridges. The rotation causes the contraction of the muscle.

2.4.2. Types of Muscle Contraction

Muscle has several types of contraction. These include twitch, isotonic and isometric and tetanus.

Twitch: This is a momentary contraction of muscle in response to a single stimulus. It is the simplest type of recordable muscle contraction.

Isotonic/Isometric: In this case a muscle contracts, becoming shorter. This results in the force or tension remaining constant as the muscle moves. For example, when you lift a weight, your muscles contract and move your arm, which pulls the weight. In contrast an isometric contraction occurs when muscle develops tension but the muscle fibres remain the same length. This is illustrated by pulling against an immovable object.

Tetanus: This results when muscle receives a stimulus at a rapid rate. It does not have time to relax before each contraction. An example of this type of contraction is seen in lock-jaw, where the muscle cannot relax due to the rate of nervous stimulus it is receiving.

Myograms: During contraction the electrical potential generated within the fibres can be recorded via external electrodes. The resulting electrical activity can be plotted on a chart. These myograms can be used to analyse various muscle contractions, both normal and abnormal.

2.4.3. Smooth Muscle

Smooth muscle tissue is so called because it does not have striations and therefore appears smooth under a microscope. It is also called involuntary because it is controlled by the autonomic nervous system. Unlike skeletal muscle, it is not attached to bone. It is found within various systems within the human body, for example the circulatory, the digestive and respiratory. Its main difference from skeletal muscle is that its contraction and relaxation are slower. Also, it has a rhythmic action which makes it ideal for the gastro-intestinal system. The rhythmic action pushes food along the stomach and intestines.

2.4.4. Cardiac Muscle

Cardiac muscle, as the name implies, is found only in the heart. Under a microscope the fibres have a similar appearance to skeletal muscle. However, the fibres are attached to each other via a specialised junction called an 'intercalated disc'. The main difference between skeletal and cardiac muscle is that cardiac muscle has the ability to contract rhythmically on its own without the need for external stimulation. This of course is of high priority in order that the heart may pump for 24 hours/day. When cardiac muscle is stimulated via a motor end plate calcium ions influx into the muscle fibres. This results in contraction of the cardiac muscle. The intercalated discs help synchronise the contraction of the fibres. Without this synchronisation the heart fibres may contract independently, thus greatly reducing the effectiveness of the muscle in pumping the blood around the body.

2.4.5. Muscle Mechanics

Movement of the skeletal structure is achieved via muscle. Skeletal muscles are classified according to the types of movement that they can perform. For simplicity, there are basically two types of muscle action – flexion and extension. Examples of flexion and extension are seen in Figure 2.9. The overall muscular system of the human body can be seen in Figures 2.10 and 2.11.

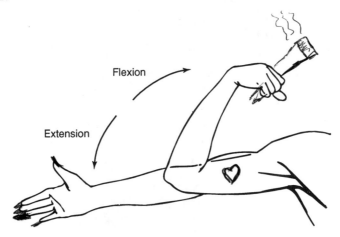

Flexion

Extension

Figure 2.9 Flexion and extension

Most body movement, even to perform such simple functions as extension or flexion, involves complex interactions of several muscles or muscle groups. This may involve one muscle antagonising another in order to achieve a specific function. The production of movement of the skeletal system involves four mechanisms – agonist, antagonist, synogists and fixators.

Agonist is a muscle that is primarily responsible for producing a movement. An antagonist opposes the movement of the prime mover. The specific contraction or relaxation of the antagonist working in co-operation with the agonist helps to produce smooth movements.

The synogist groups of muscles complement the action of the prime mover. The fixator muscles provide a stable base for the action of a prime mover – for example muscles that steady the proximal end of an arm, while the actual movement takes place in the hand.

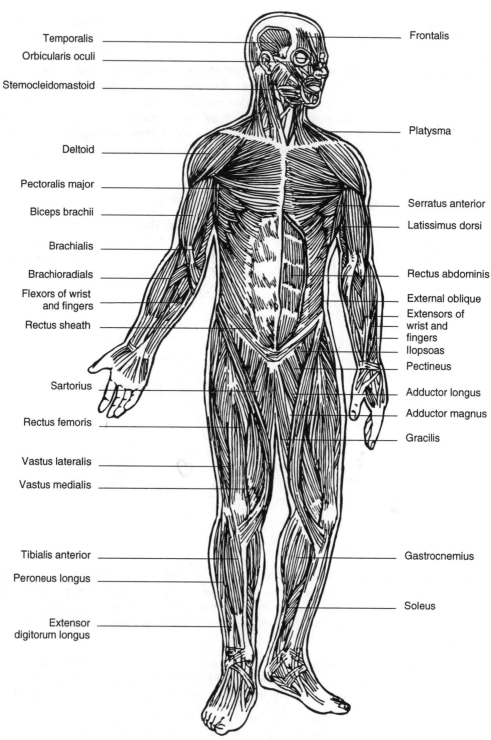

Figure 2.10 Anterior muscles of the body

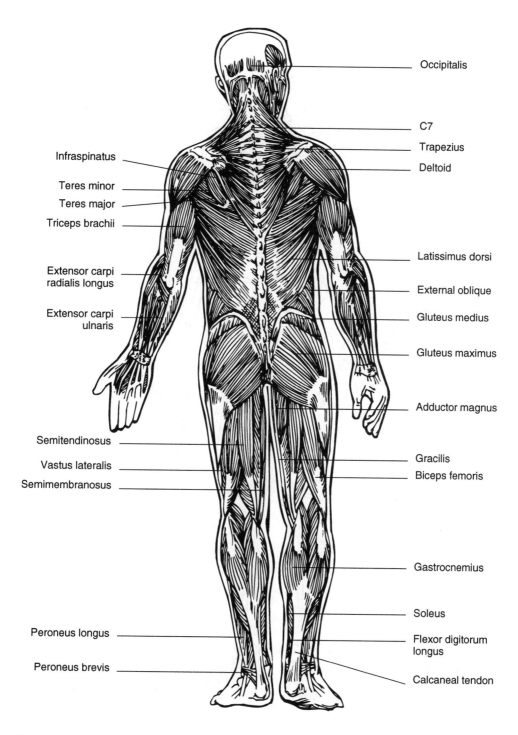

Occipitalis

C7

Trapezius

Deltoid

Infraspinatus

Teres minor

Teres major

Triceps brachii

Latissimus dorsi

Extensor carpi
radialis longus

External oblique

Gluteus medius

Extensor carpi
ulnaris

Gluteus maximus

Adductor magnus

Semitendinosus

Vastus lateralis

Semimembranosus

Gracilis

Biceps femoris

Gastrocnemius

Soleus

Peroneus longus

Flexor digitorum
longus

Peroneus brevis

Calcaneal tendon

Figure 2.11 Posterior muscles of the body

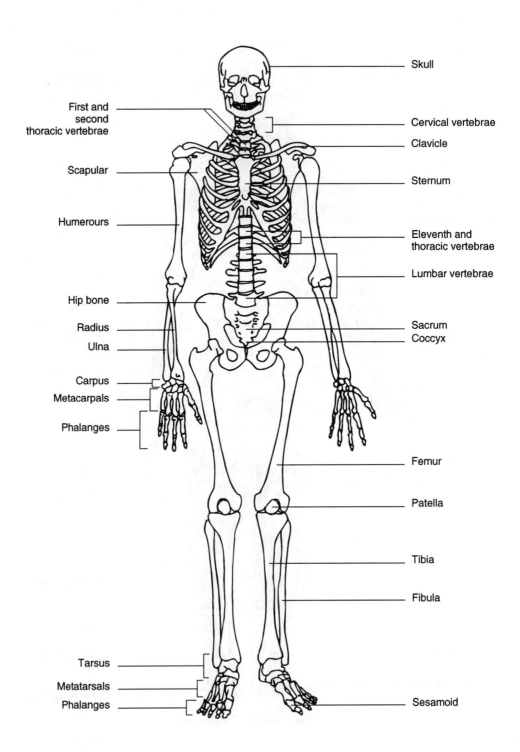

Figure 2.12 Human skeletal system

All four of these muscle groups work together with an overall objective of producing smooth movement of the skeletal structure.

Muscle is usually attached to a bone by a tendon – this is a thick cord of connective tissue comprising collagen fibres. When muscle contracts, one bone remains stationary, whilst the bone at the other end of the muscle moves. The end of the muscle that is attached to the bone that remains stationary is commonly called 'the origin', whilst the other attachment to the moving bone is called 'the insertion'.

2.5. Skeletal System

The adult skeleton consists of 206 different bones. However, it is common to find an individual with an extra rib or an additional bone in the hands or feet. Shown in Figure 2.12 is the adult human skeleton. Bone is a composite material consisting of different substances interconnected in such a way as to produce a material with outstanding mechanical properties. It consists of a matrix of an organic material, collagen, and a crystalline salts, called hydroxyapetite.

There are two types of bone – cortical (or compact) and cancellous (trabecullar). Cortical bone is a hard dense material visible on the bone's surface. Due to its appearance it is often called compact bone. Cancellous bone exists within the shell of the cortical bone (Figure 2.13). Cancellous bone is often referred to as spongy bone, as it consists of widely spaced interconnecting fibre columns called trabecullar. The centre of a long bone is filled with marrow, and this area is called the medullary cavity. It has an important role in producing blood cells during childhood. The two ends of a human long bone are called the 'epiphysis', while the mid region is referred to as the 'diaphysis'.

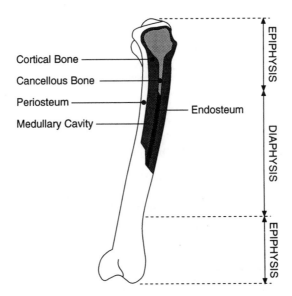

Figure 2.13 Long bone structure

Articulation of the skeletal systems occurs via joints. These joints are classified according to their movement. In hinge joints, as the name implies, movement occurs similar to that on hinges of the lid of a box. For pivot joints, the best example is the skull rotating on a peg, attached to the vertebra. Finally there are ball and socket joints, a typical example of which is found in the hip, in which the head of the femur articulates with the socket of the assetablum.

Most major joints are encapsulated and lubricated by synovial fluid. A typical example is the hip joint shown in Figure 2.14.

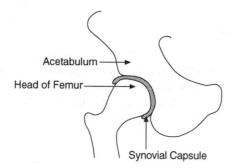

Acetabulum

Head of Femur

Synovial Capsule

Figure 2.14 Hip joint

2.6. The Nervous System

2.6.1. Anatomy

The human body reacts to a number of stimuli, both internally and externally. For example, if the hand touches a flame from a cooker, the response would be to pull the hand away as quickly as possible. The mechanism to achieve this response is controlled via the nervous system. Impulses travel from the tips of the fingers along nerves to the brain. The information is processed and the response organised. This results in the hand being pulled away from the flame using the muscular system.

The nervous system is also responsible in regulating the internal organs of the body. This is in order that homeostasis can be achieved with minimal disturbance to body function. The signals that travel along the nervous system result from electrical impulses and neuro transmitters that communicate with another body tissue, for example muscle.

For convenience, the nervous system is split into two sections, but it is important to stress that both these networks communicate with each other in order to achieve an overall steady state for the body. The two systems are termed Central and Peripheral.

The central nervous system consists of the brain and the spinal cord and can be thought of as a central processing component of the overall nervous system.

The peripheral nervous system consists of nerve cells and their fibres that emerge from the brain and spinal cord and communicate with the rest of the body. There are two types of nerve cells within the peripheral system – the afferent, or sensory nerves, which carry nerve impulses from the sensory receptors in the body to the central nervous system; and the

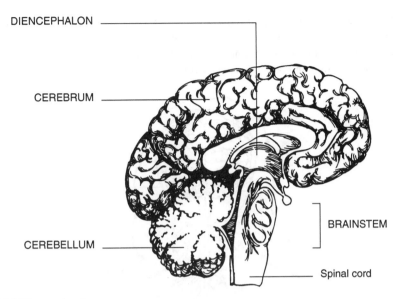

Figure 2.15 Human brain

efferent, or motor nerve cells which convey information away from the central nervous system to the effectors. These include muscles and body organs.

The highest centre of the nervous system is the brain. It has four major sub-divisions; the brain stem, the cerebellum, cerebrum and the diencephalon. The location in the brain of these various divisions is seen in Figure 2.15. Each is concerned with a specific function of the human body. The brain stem relays messages between the spinal cord and the brain. It helps control the heart rate, respiratory rate, blood pressure and is involved with hearing, taste and other senses. The cerebellum is concerned with co-ordination for skeletal muscle movement. The cerebrum concentrates on voluntary movements, and co-ordinates mental activity. The diencephalon connects the mid brain with the cerebral hemispheres. Within its area it has the control of all sensory information, except smell, and relays this information to the cerebrum. Other areas within the diencephalon control the autonomic nervous system, regulate body heat, water balance, sleep/wake patterns, food intake and behavioural responses associated with emotions.

The human brain is mostly water; about 75% in the adult. It has a consistency similar to that of set jelly. The brain is protected by the skull. It floats in a solution called the cerebrospinal fluid and is encased in three layers of tissue called the cranial meninges - the inflammation of which is termed meningitis. The brain is very well protected from the injury that could be caused by chemical compounds. Substances can only enter the brain via the blood brain barrier. The capillaries within the brain have walls that are highly impermeable and therefore prevent toxic substances causing damage to the brain. Without this protection the delicate neurons could easily be damaged.

The brain is connected to the spinal cord via the brain stem. The spinal cord extends from the skull to the lumbar region of the human back. Presented in Figure 2.16 is the distribution of the nerves from the spinal cord. Similar to the brain, the spinal cord is bathed in cerebrospinal fluid. The cord and the cerebrospinal fluid is contained within a ringed sheath called the duramatter. All these structures are contained within the vertebral column.

The vertebral column is made up of individual vertebra that are separated from each other by annular intervertebral discs. These discs have similar consistency to rubber and act as shock absorbers for the vertebral column. Each vertebra has a canal from which the spinal nerve can leave the spinal column and become a peripheral nerve. Figure 2.17 illustrates the function of

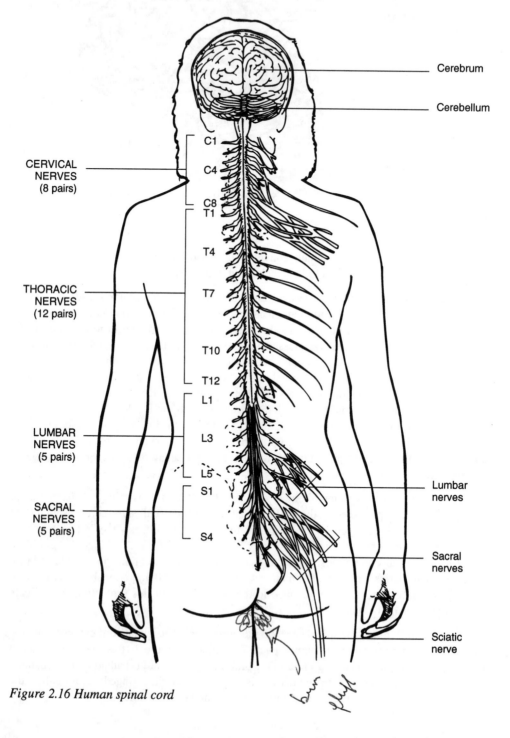

Figure 2.16 Human spinal cord

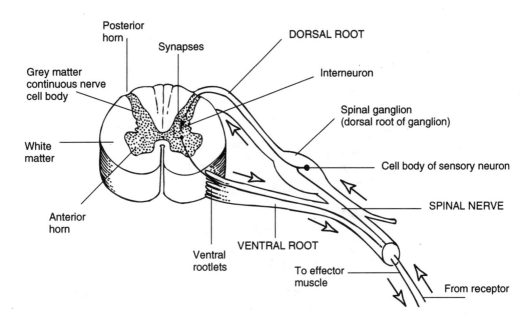

Figure 2.17 Human peripheral nerve
Derived from Carola *et al.*, 1990

a peripheral nerve. It transmits sensory information to the spinal cord, from which information can either be transmitted to the higher nervous system, the brain, for interpretation and action, or can be acted on directly within the spinal cord and the information sent back down the ventral route to initiate the response. This latter action is best illustrated by the simple reflex arc, illustrated in Figure 2.18.

If the spinal cord is injured, the resulting disability is related to the level of the injury. Injuries of the spinal cord nearer the brain result in larger loss of function compared to injuries lower down the cord. Illustrated in Figure 2.19 are two types of paralysis that can occur due to transection of the cord.

Paraplegia is the loss of motor and sensory functions in the legs. This results if the cord is injured in the thoracic or upper lumbar region. Quadriplegia involves paralysis of all four limbs and occurs from injury at the cervical region. Hemiplegia results in the paralysis of the upper and lower limbs on one side of the body. This occurs due to the rupture of an artery within the brain. Due to the architecture of the connections between the right and left hand side of the brain, damage to the right hand side of the brain would result in hemiplegia in the opposite side.

2.6.2. Neurons

The nervous system contains over one hundred billion nerve cells, or Neurons. They are specialised cells which enable the transmission of impulses from one part of the body to another via the central nervous system.

Neurons have two properties; excitability, or the ability to respond to stimuli; and conductivity, the ability to conduct a signal. A neuron is shown diagrammatically in Figure 2.20.

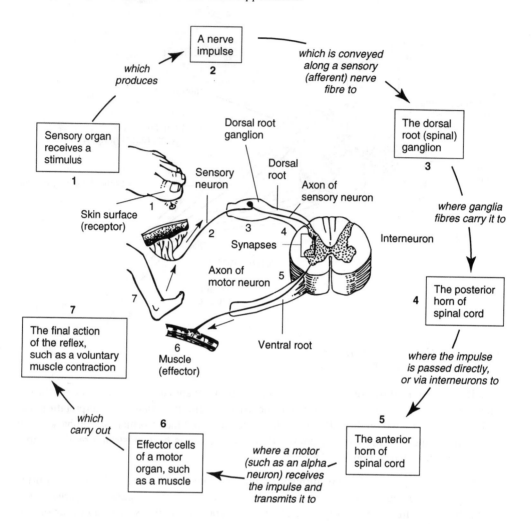

Figure 2.18 Nerve reflex arc
Derived from Carola *et al.*, 1990

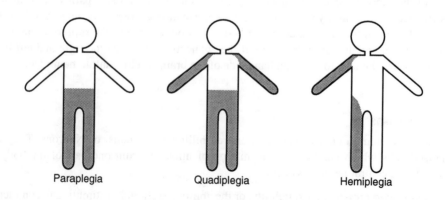

Figure 2.19 Types of paralysis due to transection of the spinal cord

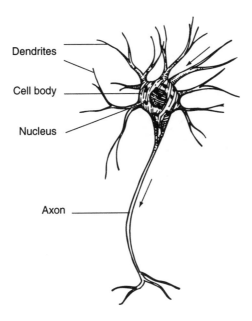

Dendrites

Cell body

Nucleus

Axon

Figure 2.20 Neuron

Dendrites conduct information towards the cell body. The axon transmits the information away from the cell body to another nerve body tissue. Some axons have a sheath which is called myelin. The myelin sheath is segmented and interrupted at regular intervals by gaps called neurofibral nodes. The gaps have an important function in the transmission of impulses along the axon. This is achieved via neurotransmitters. Unmyelinated nerve fibres can be found in the peripheral nervous system. Unlike the myelinated fibres they tend to conduct at a slower speed.

2.6.3. Physiology of Neurons

Neurons transmit information via electrical pulses. Similar to all other body cells, transmission depends upon the difference in potential across the membrane of the cell wall. With reference to Figure 2.21, a resting neuron, is said to be polarised, meaning that the inside of the axon is negatively charged with relation to its outside environment. The difference in the electrical charge is called the potential difference. Normally the resting membrane potential is −70 mV. This is due to the unequal distribution of potassium ions within the axon and sodium ions outside the axon membrane. There are more positively charged ions outside compared to within the axon.

Figure 2.22 shows the sodium/potassium pump that is found in the axon membrane. This pump is powered by ATP and transports three sodium ions out of the cell for every two potassium ions that enter the cell.

In addition to the pump the axon membrane is selectively permeable to sodium/potassium through voltage gates, known as open ion channels. These come into operation when the concentration of sodium or potassium becomes so high on either side that the channels open up to re-establish the distribution of the ions in the neuron at its resting state (−70 mV).

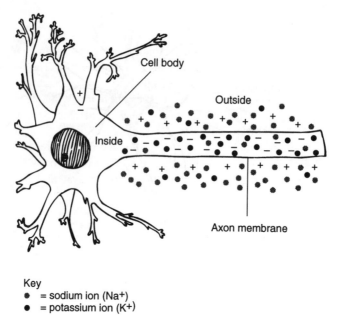

Cell body

Outside

Inside

Axon membrane

Key
● = sodium ion (Na+)
● = potassium ion (K+)

Figure 2.21 Ions associated with neuron
Derived from Carola et al., 1990

2.6.4. The Mechanism of Nerve Impulses

The process of conduction differs slightly between unmyelinated and myelinated fibres. For unmyelinated fibres the stimulus has to be strong enough to initiate conduction. The opening of ion channels starts the process called *depolarisation*.

Once an area of the axon is depolarised it stimulates the adjacent area and the action potential travels down the axon. After depolarisation the original balance of sodium on the outside of the axon and potassium inside is re-stored by the action of the sodium/potassium pumps. The membrane is now *re-polarised*.

There is a finite period whereby it is impossible to stimulate the axon in order to generate an action potential. This is called the refractory period and can last anything from 0.5 to 1 ms. A minimum stimulus is necessary to initiate an action potential. An increase in the intensity of the stimulus does not increase the strength of the impulse. This is called an all or none principle. In myelinated fibres the passage of the impulse is speeded up. This is because the myelin sheath around the axon acts as an insulator and the impulses jump from one neurofibral node to another. The speed of conduction in unmyelinated fibres ranged from 0.7 to 2.3 metres/second, compared with 120 metres/second in myelinated fibres.

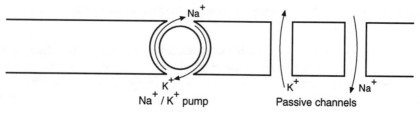

Na^+

K^+

Na^+ / K^+ pump

K^+ Na^+

Passive channels

Figure 2.22 The sodium/potassium pump

2.6.5. The Autonomic Nervous System

A continuation of the nervous system is the Autonomic nervous system, which is responsible in maintaining the body's homeostasis without conscious effort. The autonomic nervous system is divided into sympathetic and para-sympathetic. The responsibility of each of these divisions is shown in Tables 2.2 and 2.3. The best example involving the autonomic nervous system is the 'Flight or fight' reaction. Most people have experienced this in the form of fear. The body automatically sets itself up for two responses – either to 'confront' the stimuli, or run away. The decision on which to do is analysed on a conscious level. It is obvious from looking at the roles of these divisions that the homeostasis of the body would be extremely difficult, if not impossible, to achieve without this important system. Failure of any of these effects would be a life threatening condition.

Table 2.2 Sympathetic System – Neurotransmitter Noradrenaline

Action	Effects
radial muscle of pupil (+)	dilation of pupil
salivary glands (+)	secretion of thick saliva
blood vessels { (+)	vasoconstriction
{ (–)	vasodilation
heart (+)	rate and force increased
lung (airways (–)	bronchodilation
gut wall (-)	decrease in
gut sphincters (+)	motility and tone
	glycogenolysis
liver (+)	gluconeogenesis (glucose release into blood)
spleen (+)	capsule contracts
adrenal medulla (+) ⟶	ADRENALINE
bladder	relaxation
detrusor (-)	contraction
sphincter (+)	
uterus { (+)	contraction or
{ (–)	relaxation
vas deferens (+)	
seminal	ejaculation
vesicles (+)	
	muscarinici
sweat glands (+)	sweating
pilomotor muscles	pilo-erection (hairs stand on end)

Table 2.3 Parasympathetic System – Neurotransmitter Acetylcholine

Action	Effects
(+) lacrimal gland	tear secretion
(+) circular muscle of iris	constriction of pupil
(+) ciliary muscle	accommodation for near vision
(+) salivary glands	much secretion of watery saliva
(–) heart	rate and force reduced
(+) lung airways bronchosecretion	bronchoconstriction
(+) gut wall	
(–) gut sphincters	increase in motility and tone
(+) gut secretions increase in	
(+) pancreas endocrine secretion	exocrine and
(+) bladder detrusor	micturition
(–) sphincter	
(+) rectum	defaecation
(+) penis venous sphincters contracted	erection

2.7. The Cardio-Vascular System

The centre of the cardio-vascular system is the heart. The heart can be considered as a four chambered pump. It receives oxygen deficient blood from the body; sends it to get a fresh supply of oxygen from the lungs; then pumps this oxygen rich blood back round the body. It has approximately 70 beats per minute and 100,000 per day. Over 70 years the human heart pumps 2.5 billion times. Its size is that approximately of the clenched fist of its owner and it weighs anything between 200 and 400 grams, depending upon the sex of the individual. It is located in the centre of the chest, with two thirds of its body to the left of the mid line.

Heart muscle is of a special variety, termed cardiac. Due to the inter-collated discs, the cells act together in order to beat synchronously to achieve the aim of pumping the blood around the body. The physiology of the action potential within the cells is similar to that of the nerves. The anatomical structure of the heart is shown in Figure 2.23. De-oxygenated blood returns from the body via the veins into the right atrium. The right atrium contracts, sending the blood into the right ventricle. The one-way valve enables the blood, on the contraction of the right ventricle, to be expelled to the lungs, where it is oxygenated (pulmonary system). The returning oxygenated blood is fed into the left atrium, and then into the left ventricle. On contraction of the left ventricle, again via a one-way valve, the blood is sent to the various parts of the body via blood vessels (Figure 2.24). The systemic/pulmonary cardiac cycle is shown in Figure 2.25. The whole cycle is repeated 70 times per minute.

The contraction of the cardiac muscle is initiated by a built-in pacemaker that is independent of the central nervous system. With reference to Figure 2.26, the specialised nervous tissue in the right atrium is called the sino atrial node; it is responsible for initiating contraction. The

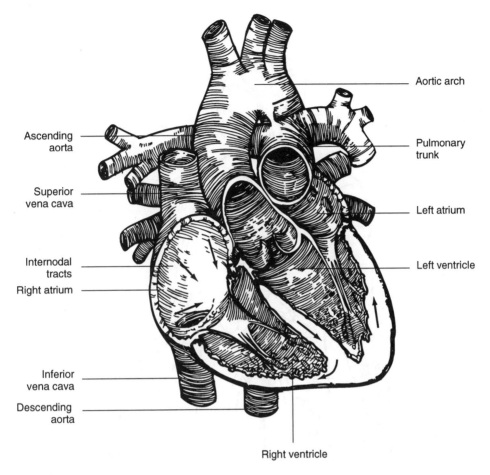

Figure 2.23 Human heart

signals are passed down various nervous pathways to the atrio-ventricular node. This causes the two atria to contract. The nervous signal then travels down the atrio-ventricular bundles to initiate the contraction of the ventricles. The transmission of the various impulses along these pathways gives off an electrical signal. It is the measurement of these signals that produce the electro-cardiograph (ECG)(Figure 2.27).

The P region of the electro-cardiograph represents atrial contraction. The ventricular contractions are represented by the QRS wave, whilst the T waveform is ventricular relaxation. Typical times for the duration of the various complexes are shown in Table 2.4. Recording of these signals is obtained by placing electrodes on various parts of the body. These are shown in Figure 2.28. Other than their own specialised cells to conduct the nerve impulses, the heart receives other nerve signals. These come mainly from the sympathetic and para-sympathetic autonomic nervous system. The sympathetic system, when stimulated, tends to speed up the heart, while the parasympathetic system tends to slow the heart rate down. If for some reason the mechanism for transmitting the nervous signals from the atrium to the ventricles is disrupted, then the heart must be paced externally. This can be achieved by an electronic device called the pacemaker. This device feeds an electrical current via a wire into the right ventricle. This passes an impulse at a rate of approximately seventy per minute.

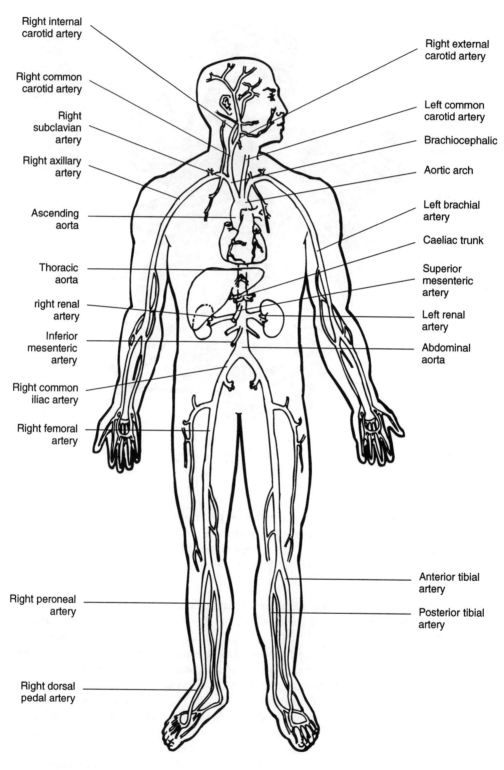

Figure 2.24a Arterial system

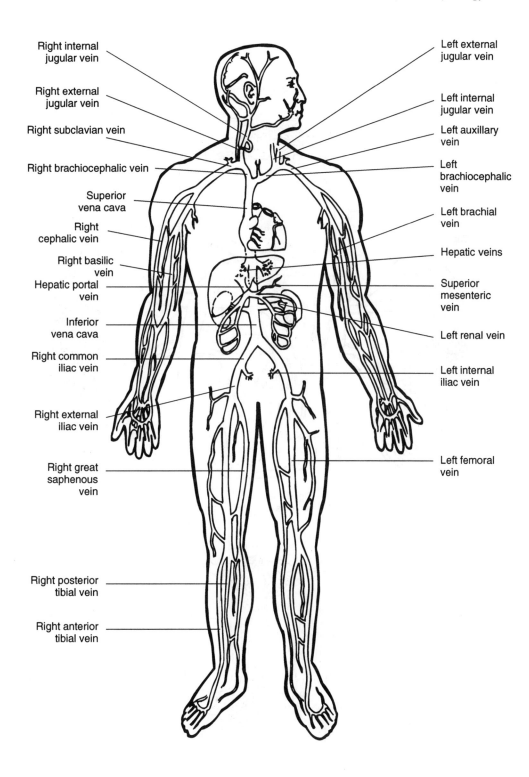

Figure 2.24b Venous system

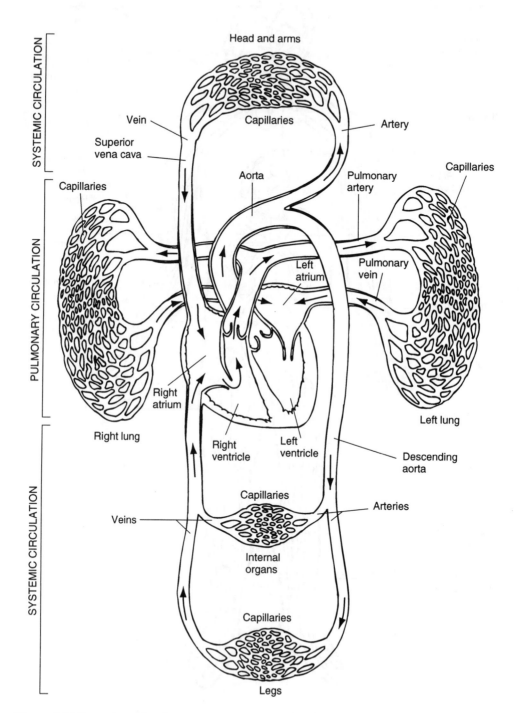

Figure 2.25 Systemic and pulmonary system
Derived from Carola et al., *1990*

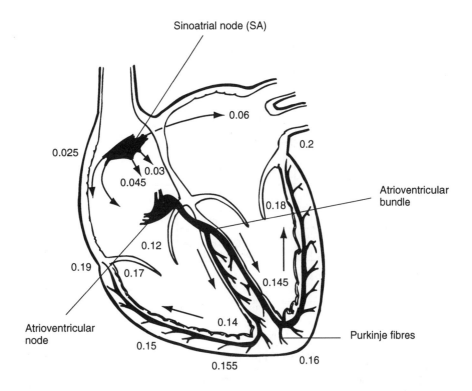

Figure 2.26 Nerve conduction times within the heart
Derived from Carola et al., 1990

Table 2.4 Transmission times in the heart

ECG Event	Range of duration (seconds)
P wave	0.06 – 0.11
P–R segment	0.06 – 0.10
(wave)	
P–R interval	0.12 – 0.21
(onset of P wave to onset of QRS complex)	
QRS complex	0.03 – 0.10
(wave and interval)	
S–T segment	0.10 –.0.15
(wave) (end of QRS complex to onset of T wave)	
T wave	Varies
S–T interval	0.23 – 0.39
(end of QRS complex to end of T wave)	
Q–T interval	0.26 – 0.49
(onset of QRS complex to end of T wave)	

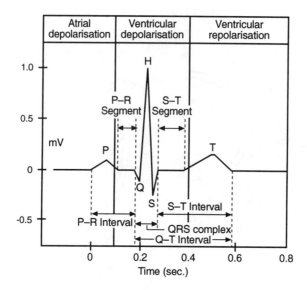

Figure 2.27 A typical ECG

2.7.1. Measurement of Blood Pressure

When the heart contracts, it circulates blood throughout the body. The pressure of the blood against the wall is defined as the blood pressure. Its unit of measurement is millimetres of mercury (mmHg). When the ventricles contract, the pressure of the blood entering the arterial system is termed systolic. The diastolic pressure corresponds to the relaxation of the ventricle. The difference between these two pressures is termed the blood pressure (systolic/diastolic). A normal young adult's blood pressure is 120/80 mmHg. If the blood pressure is considerably higher then the patient is termed to be hypertensive. Blood pressure varies with age. The systolic pressure of a new-born baby may only be 40, but for a 60 year old man it could be 140 mmHg. Causes of abnormal rises in blood pressure are numerous. Blood pressure rises temporarily during exercise or stressful conditions and a systolic reading of 200 mmHg would not be considered abnormal under these circumstances.

2.8. Respiratory System

The body requires a constant supply of oxygen in order to live. The respiratory system delivers oxygen to various tissues and removes metabolic waste from these tissues via the blood. The respiratory tract is shown in Figure 2.29.

Breathing requires the continual work of the muscles in the chest wall. Contraction of the diaphragm and external intercostal muscles expands the lungs' volume and air enters the lungs. For expiration, the external intercostal muscles and the diaphragm relax, allowing the lung volume to contract. This is accompanied by the contraction of abdominal muscles and the elasticity of the lungs.

We return to a discussion of measurement of cardio-vascular function and the control of certain of its disorders in Chapter 4.

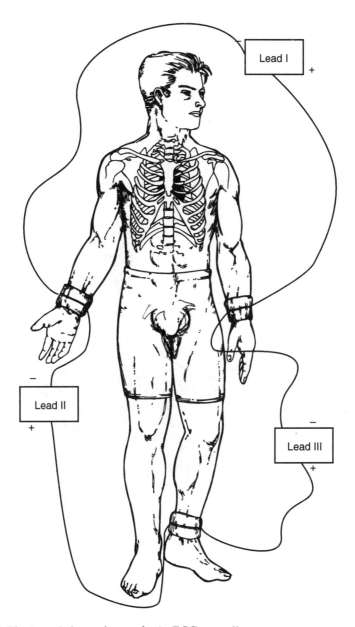

Figure 2.28 Placing of electrodes to obtain ECG recording

2.8.1. Volumes of Air in the Lung

With reference to Figure 2.30, pulmonary ventilation can be broken down into various volumes and capacities. These measurements are obtained using a respirometer. During normal breathing at rest, both men and women inhale and exhale about 0.5 litre with each breath – this is termed the tidal volume.

The composition of respiratory gases entering and leaving the lungs is shown in Table 2.5.

Table 2.5 Composition of main respiratory gases entering and leaving lungs (standard atmospheric pressure, young adult male at rest)

	Oxygen volume %	Carbon dioxide volume %	Nitrogen volume %
Inspired air	21	0.04	78.0
Expired air	16	4.0	79.2
Alveolar air	14	5.5	79.1

Percentages do not add up to 100 because water is also a component of air.

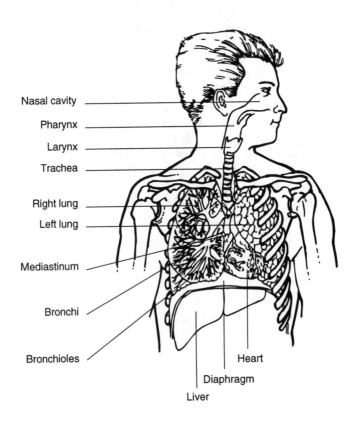

Nasal cavity
Pharynx
Larynx
Trachea
Right lung
Left lung
Mediastinum
Bronchi
Bronchioles
Heart
Diaphragm
Liver

Figure 2.29 Respiratory tract

2.8.2. Diffusion of Gases

The terminal branches in the lung are called the alveoli. Next to the alveoli are small capillaries. Oxygen and carbon dioxide are transported across the alveoli membrane wall. Various factors affect the diffusion of oxygen and carbon dioxide across the alveoli capillary membrane. These include the partial pressure from either side of the membrane, the surface area, the thickness of the membrane, and solubility and size of the molecules.

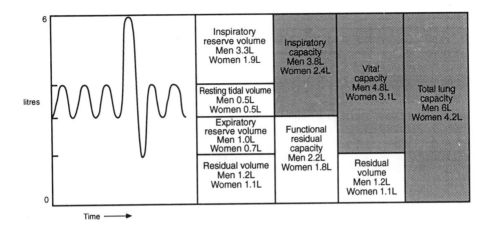

Figure 2.30 Various pulmonary volumes and capacities
Derived from Carola et al., 1990

The inspired oxygen transfers across the alveoli membrane to the red blood cells in the capillaries. Oxygen attaches itself to the haemoglobin, whilst carbon dioxide is released from the haemoglobin and travels in the reverse direction to the alveoli. The carbon dioxide is then expired as waste through the respiratory system. Similarly, at the tissue, the oxygen is released from the red blood cells and is transported across the tissue membrane to the tissue. Carbon dioxide travels in the opposite direction.

The transportation of oxygen and carbon dioxide in the red blood cells depends upon the concentration of a protein called haemoglobin. Haemoglobin has a high affinity for oxygen and therefore is a necessary component in the transfer of oxygen around the human body.

2.8.3. The Control of Breathing

The rate and depth of breathing can be controlled consciously but generally it is regulated via involuntary nerve impulses. This involuntary process is mediated via the medullary area of the central nervous system.

3

Physics

3.1. The nature of ionising radiation

Ionising radiation is the term used to describe highly energetic particles or waves which when they collide with atoms cause the target atoms to receive significant kinetic energy. This energy may cause inelastic collisions when the target atom absorbs a proportion of the energy and is placed into a higher energy state. Alternatively the incident energy may be divided between the source and the target, in which both are displaced with energies whose sum is the total incident energy. The specific mechanism which occurs is dependent on the nature of the incident radiation, the target and the incident energy level. Ionising radiation is produced by one of several phenomena.

1. Cosmic radiation, which is mainly due to extra-terrestrial nuclear reactions. The nuclear processes which take place in the sun and the stars result in a small but measurable flux of high energy radiation which reaches and in some cases traverses the earth. This radiation is found (UNSC, 1982) to give rise to somewhat more than one tenth of our annual natural background radiation dose.

2. Nuclear decay of unstable elements on the earth. There are rocks in the earth's crust containing unstable elements which undergo nuclear decay. This process gives rise to a small amount of nuclear radiation, but as part of the decay, certain of the child products are also radioactive, and these in turn give rise to radiation. In particular, there are significant doses due to the decay of radon-222 and radon-220 through absorption through the lungs. Radon is locally concentrated owing to the types of subsoil and building material used.

3. Artificial production of ionising radiation from high energy sources. If an energetic electron collides with an atom, it may give up its energy inelastically and produce high energy photons, some of which are in the X ray region.

3.1.1. Sources of X rays

A medical X ray tube (Figure 3.1) is built from a vacuum tube with a heated cathode which emits electrons. These are accelerated by a high electric potential (of up to around 300 kV) towards a target anode. The anode is built from a metal with a high atomic number to provide the best efficiency of conversion of the incident electron energy into photons. Nonetheless the typical efficiency is only about 0.7%. With typical tube currents of 10 to 500 mA, instantaneous input powers up to 100 kW may be used. As a result, there is a significant problem of

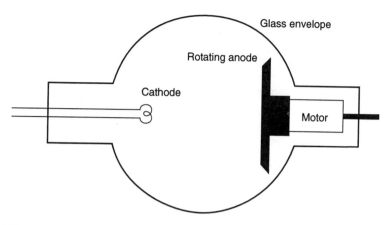

Figure 3.1 X ray tube

anode heating. To reduce this problem, the anode is normally rotated at high speed (about 3600 rpm), and is normally made with a metal which has a high melting point. Frequently the anode target layer is relatively thin and is backed by copper to improve thermal conduction. In spite of these precautions, localised temperatures on the anode may reach around 2500°C.

A typical tube is about 8–10 cm in diameter and 15–20 cm in length.

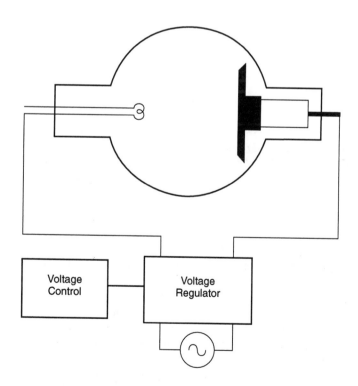

Figure 3.2 Simplified circuit for driving X ray tube

Figure 3.2 shows a simplified schematic for an X ray power circuit. In practice, the tube is driven in modern sets from sophisticated supplies. In some forms of radiography which are outlined in Chapter 6 X ray pulses are used over a prolonged period to obtain images. Frequently these require pulse durations of milliseconds. It is crucial in most of these cases that the high voltage power supply is stable to ensure that radiation of the required spectrum is produced.

3.1.1.1. X Ray Spectra

The spectral emission of an X ray tube is shown in Figure 3.3.

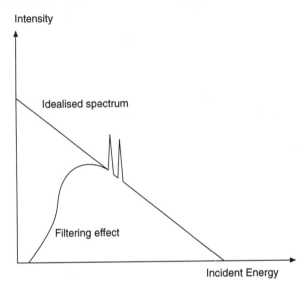

Figure 3.3 Form of X ray spectrum

The radiation emitted by an X ray tube is primarily due to *Bremsstrahlung* – the effect of the deceleration of the high energy electron beam by the target material of the X ray tube. The incident electrons collide with the nuclei in the target, and some result in inelastic collisions in which a photon is emitted. The peak photon energy is controlled by the peak excitation potential used to drive the tube. For a thin target, the radiant photon energy is of a uniform distribution.

When a thick target is used, as is normally the case for diagnostic applications, the incident electrons may undergo a number of collisions in order to lose their energy. Collisions may therefore take place throughout the depth of the target anode. Through the thickness of the target, there is a progressive reduction of the mean incident energy. The result would then be a spectrum decreasing linearly from zero to the peak energy. However, the target material also absorbs a proportion of the generated X rays, preferentially at the low energy end of the spectrum. There are also strong spectral lines produced in the X ray spectrum as a result of the displacement of inner shell electrons by incident electrons. The form of the resultant spectrum is as shown in Figure 3.3.

Protection against the low energy components of this spectrum may be enhanced by the use of filters. These are used when the low energy components would not penetrate the area of

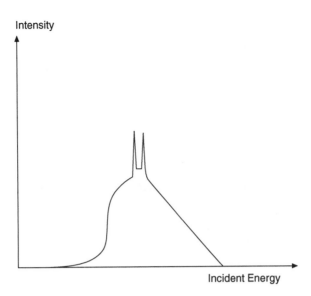

Figure 3.4 Filtered X ray spectrum

interest adequately, and therefore simply cause potential problems due to an unnecessary radiation dose. Low energy components may of course be used alone when they are adequate to pass through small volumes of tissue in applications such as mammography. The filtered spectrum is shown in Figure 3.4.

3.1.2. Radioactive decay

The nuclei of large atoms tend to be unstable and susceptible to decay by one of several mechanisms.

1. Beta emission, in which an electron is emitted from the nucleus of the atom, approximately retaining the atomic weight of the atom, but incrementing the atomic number.

2. Alpha emission is due to the release of the nucleus of a helium atom from the decaying nucleus. Alpha particles are released with energies in the range 4–8 MeV, but readily lose their energy in collisions with other matter.

3. Neutrons are emitted when a nucleus reduces its mass, but does not change its atomic number. Neutrons are emitted either spontaneously or as a result of an unstable atom absorbing a colliding neutron and then splitting into two much smaller parts together with the release of further neutrons.

4. Gamma radiation is the emission of high energy photons from unstable atomic nuclei. This occurs frequently following either of the previously mentioned forms of decay, which often leave the resulting atomic nucleus in a metastable state.

Radioactive decay is a probabilistic process: at any time there is a constant probability that any atom will spontaneously decay by one of these processes. The decay is not immediate, since there is an energy well which must be traversed before it may take place: the probability of the decay event taking place is related to the depth of the well. Thus for a population of N atoms the radioactivity Q is given by

$$Q = -\lambda N = dN/dt \tag{1}$$

The decay constant is related to the 'half-life' of the nuclide by:

$$T_{1/2} = (\ln 2)/\lambda \tag{2}$$

Radioactive decomposition of large nuclei takes place in a series of steps. For example, both uranium and radium are present in the earth's crust in significant quantities. They decay through a series of energy reducing steps until they ultimately become lead.

3.2. Physics of radiation absorption, types of collision

There are several characteristic modes of collision between ionising radiation and matter: the mechanism depends on the form and energy of the radiation, and the matter on which it is incident. For our purposes, the following are the most significant.

1. The Photoelectric Effect, in which an incident photon gives up all its energy to a planetary electron. The electron is then emitted from the atom with the kinetic energy it received from the incident photon less the energy used to remove the electron from the atomic nucleus. Clearly this process can only occur when the incident photon energy is greater than the electron's binding energy. The probability of this interaction decreases as the photon energy increases.

   The result of the electron loss is to ionise the atom, and if one of the inner shell electrons is removed, to leave the atom in an excited state. The atom leaves the excited state when an electron descends from an outer orbit to replace the vacancy, and a photon may be emitted, again having X ray energy. The photoelectric effect is the predominant means of absorption of ionising radiation when the incident photon energy is low.

2. The Compton Effect occurs when an incident electron collides with a free electron. The free electron receives part of the energy of the incident photon, and a photon of longer wavelength is scattered. This effect is responsible for the production of lower energy photons which are detected in nuclear medicine systems (see Chapter 6). Their reduced energy means that they may subsequently be recognised by the detection system and largely eliminated from the resulting image.

3. Pair Production occurs when a highly energetic photon interacts with an atomic nucleus. Its energy is converted into the mass and kinetic energy of a pair of positive and negative electrons. This process may only occur once the incident energy exceeds the mass equivalent of two electrons (i.e. $2mc^2 = 1.02$ MeV).

4. Neutron Collisions result in a wide range of recoil phenomena: in the simplest case, the target atomic nucleus receives some kinetic energy from the incident neutron, which is itself deflected with a reduced energy. Other forms of collision take place, including the capture of incident neutrons leaving the atom in an excited state from which it must relax by further emission of energy.

For a fuller description of these effects, the reader is referred to specialist texts, such as Greening (1981).

3.3. Radiation Measurement and Dosimetry

3.3.1. Dosimetric Units

We must firstly consider what is meant by radiation dose. Ionising radiation incident on matter interacts with it, possibly by one of the means outlined above. In doing so, it releases at least part of its energy to the matter.

As a simple measure, we may look at the rate of arrival of radiation incident on a sphere of cross section da. The *fluence* is

$$\Phi = dN/da$$

where dN is the number of incident photons or particles, and the fluence rate is $\phi = d\Phi/dt$.

The unit of this measurement is therefore $m^{-2}s^{-1}$. If the energy carried by the particles is now considered, the *energy fluence rate* may be derived comparatively in units of Wm^{-2}. The spectral intensity of the incident radiation is dependent on a number of factors: it is often important to be able to assess the spectral distribution of the incident radiation.

The unit of decay activity of radionuclides was the Curie, which became standardised at $3.7 \times 10^{10} s^{-1}$. This is approximately the disintegration rate of a gram of radium. As the SI unit of rate is s^{-1}, this unit has now been superseded by the Bequerel (Bq) with unit s^{-1} when applied to radioactive decay.

The unit of *absorbed dose*, being the amount of energy absorbed by unit mass of material, was originally the Rad, or 100 erg g^{-1} (10^{-2} J kg^{-1}) of absorbed energy. This has now also been superseded by the SI unit the Gray (Gy), which is defined as 1 J kg^{-1}.

Another unit of interest relates to *exposure* to ionising photon radiation. This measure quantifies the ionisation of air as a result of incident energy. The Roentgen (R) is defined as 2.58×10^{-4} C kg^{-1}.

The term *'Dose Equivalent'* is used to denote a weighted measure of radiation dose: the weighting factor is derived from the stopping power in water for that type and energy of radiation. This measure is normally expressed in the unit Sievert (Sv) which has the same dimensions as the Gray, but is given a special name to denote its different basis.

The *'Effective Dose Equivalent'* is the measure used to denote dose equivalent when it has been adjusted to take account of the differing susceptibilities of different corporal organs to radiation. The Effective Dose Equivalent is defined as:

$$H_E = \sum w_T H_T \tag{3}$$

The weighting factors employed here vary between 0.25 for the gonads to 0.03 for bone surface, and 0.3 for the bulk of body tissue.

3.3.2 Outline of Major Dosimetric Methods

A wide range of methods exists for the measurement of radiation dose. They include fundamental methods which rely on calorimetric measurement and measurement of ionisation: these are required as standards for the assessment of other techniques. Scintillation counters, which also help to characterise the received radiation, are described in Section 6.3.

However, in practical terms, two main methods outlined below are used in monitoring individual exposures to radiation. In addition, as a basic protection, it is frequently wise to have available radiation counters when dealing with radioactive materials as they provide real time readings of the level of radiation present.

3.3.2.1. Thermoluminescence

Many crystalline materials when irradiated store electron energy in traps. These are energy wells from which the electron must be excited in order for it to return via the conduction band to a rest potential. The return of the electron to the rest state from the conduction band is accompanied by the release of a photon which may be detected by a photomultiplier. The trapping and thermoluminescent release processes are shown in Figure 3.5. If the trap state is sufficiently deep, the probability of the electron escaping spontaneously may be sufficiently low for the material to retain the electron in the excited state for a long period: it can be released by heating the material and observing the total light output. Various materials are used, but they should ideally have similar atomic number components to that of tissue if the radiation absorption characteristics are to have similar energy dependencies. The materials are used either in a powder form in capsules or alternatively embedded in a plastic matrix.

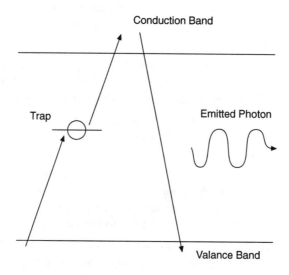

Figure 3.5 Trapping and thermoluminescence

3.3.2.2. Film Badge

The photographic film badge is a familiar and rough and readily portable transducer for the measurement of radiation dose. The film is blackened by incident radiation, although unfortunately its energy response does not closely match that of tissue. The badge holder therefore contains various metal filters which provide a degree of discrimination between different types of and energies of incident radiation. The badges worn by radiation workers are typically swapped and read out on a monthly basis to provide a continuing record of their exposure to radiation.

3.4. Outline of the Application of Radiation in Medicine – Radiology, Radiotherapy

Ionising radiation is used in medicine in two main applications. As the radiation is in some cases very energetic it is able to pass through body tissue with limited absorption. The differential absorption of radiation in different types of tissue makes it possible to obtain images of the internal structures of the body by looking at the remaining radiation if a beam of X or gamma radiation is shone through a region of the body. Absorbed doses (see section 3.3.1) from diagnostic investigations are typically around 0.1 mSv.

Additionally, radioactive substances may be injected into the body as 'labels' in biochemical materials which are designed to localise themselves to particular organs or parts of organs. The radiation emitted from the decay of these materials may be examined externally to derive an image of the organ's condition. As ionising radiation presents a significant risk of causing biological damage to tissue, if large doses of radiation may be administered to specific areas of body tissue it is possible to destroy cancerous tissue selectively and without the risks entailed with surgery. The doses involved in radiotherapy are much higher, being typically localised doses of tens of Gy delivered in smaller doses of a few Gy at intervals of several days.

3.5. Physics of NMR

Nuclear Magnetic Resonance is a physical effect which has become increasingly used in medical imaging since the 1970s. This section provides a simple outline of the physics of the NMR process. An overview of the instrumentation which is used to obtain images from this process is presented in Chapter 4.

In essence we will find that images using NMR are effectively maps of the concentration of hydrogen atoms. The images obtained are of high resolution. The display is derived from details of a subject's morphology based on factors different from those examined when conventional radiological studies are made. The examination technique has fewer apparent inherent dangers than does the use of ionising radiation, but has the serious drawback of the high capital cost of the equipment used to obtain images.

3.5.1. Precessional Motion

Probably the easiest point to start an understanding of NMR is by looking at the motion of a spinning particle in a field. Consider a child's spinning top. If it is placed spinning so that one end of its axis is pivoted, then the mass of the top acts with the earth's gravitational field and the reaction of the pivot to form a couple which tends to rotate the spinning angular momentum vector downwards (Figure 3.6). Since however angular momentum is conserved, a couple is produced which causes the top to make a precessional motion about its pivot.

Expressing this in mathematical notation, and using the symbols from the diagram (note that **bold** type refers to vector quantities), a torque is caused by gravity acting on the mass of the top:

$$\tau = \mathbf{r} \times m\mathbf{g} \tag{4}$$

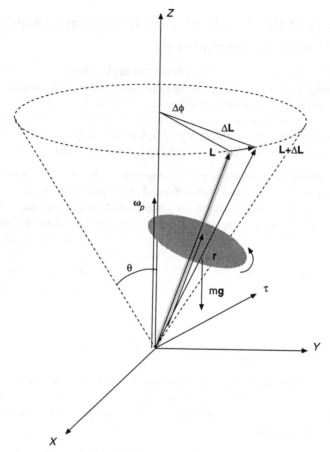

Figure 3.6 Forces acting on a gyroscope

Here the symbol × is the vector cross product. This torque acts on the gyroscope whose angular momentum is **L** to modify it, so that:

$$\tau = \frac{d\mathbf{L}}{dt}$$

(5)

In a short time t the angular momentum of the gyroscope is modified by a small amount $\Delta\mathbf{L}$ acting perpendicularly to **L**. The precessional angular velocity of the gyroscope, which is the rate at which its axis rotates about the z co-ordinate, may now be derived:

$$\omega_p = \frac{\Delta\phi}{\Delta t}$$

(6)

Since we are looking at a small change in $\Delta\mathbf{L} \ll \mathbf{L}$, the small angle $\Delta\phi$ is

$$\Delta\phi \cong \frac{\Delta\mathbf{L}}{\mathbf{L}\sin\theta} = \frac{\tau\Delta t}{\mathbf{L}\sin\theta}$$

(7)

and the precessional velocity from equation 6 above is

$$\omega_p = \frac{\Delta\phi}{\Delta t} = \frac{\tau}{\mathbf{L}\sin\theta}$$

(8)

Substituting for τ from equation 4, we obtain an expression for the magnitude of the angular velocity of the precessional motion:

$$\omega_p = {mgr}/{L} \tag{9}$$

This tells us that the precessional angular velocity is proportional to the force due to the field (mg) and inversely proportional to the body's angular momentum.

The NMR phenomenon is analogous. A spinning charge (in the simplest case a proton, the nucleus of a hydrogen atom) if placed in a magnetic field precesses about the field. The spin vector representing angular momentum may be either directed with or against the magnetic field: the two directions possible with hydrogen represent two different energy states. Evaluation of the concentration of hydrogen is undertaken by stimulating a proportion of the nuclei into the higher energy state with a radio frequency electromagnetic pulse and then examining the energy released as they decay into the lower state. The following paragraphs provide a mathematical statement of the effect so that it may be quantified.

Firstly, a rotating charge has a magnetic moment :

$$\mathbf{m} = \gamma \mathbf{I} \tag{10}$$

in which $\mathbf{m}$ is the magnetic moment, and $\mathbf{I}$ the angular momentum, and $1/\gamma$ is the gyromagnetic ratio. In classical physics, γ is $e/2m$ where e is the charge and m the mass of the particle.

If the rotating charge is placed in a magnetic field of strength $\mathbf{B}$, the field causes a torque which makes the particle's magnetic moment and, as a result, also its momentum vector, precess about the direction of the field. The rate of change of the particle's momentum then is given by

$$d\mathbf{I}/{dt} = \mathbf{m} \times \mathbf{B} \tag{11}$$

Now substitute in equation 10 to yield

$$d\mathbf{m}/{dt} = \gamma \mathbf{m} \times \mathbf{B} \tag{12}$$

In the steady state, the precession continues indefinitely with an angular velocity given by

$$\omega = -\gamma \mathbf{B} \tag{13}$$

This expression has the same form as that of the expression which we derived for a spinning top. In this case the precessional velocity is proportional to the strength of the applied magnetic field.

We now may briefly extend our view to include a quantum mechanical description of the motion. In this view, energy states and angular momentum are discrete rather than a continuum of values. In the case of a hydrogen nucleus, the permitted values of the spin quantum numbers are $\pm\frac{1}{2}$, representing spin vectors with and against the magnetic field. The respective energy states are

$$E = \pm \tfrac{1}{2} \gamma \hbar |\mathbf{B}| \tag{14}$$

where $\hbar$ is $2\pi h$ and h is Plank's constant. The separation of the levels is

$$\Delta E = \gamma \hbar |\mathbf{B}| \tag{15}$$

These expressions describe the precession of the momentum vector in terms of a fixed system of '*Laboratory Co-ordinates*'. We could instead describe the equations in terms of some other set of co-ordinates. It will turn out to be easier to understand the origin of the later expressions and visualise the processes if we transform equation 12, which is known as the *Larmor Equation*, into a rotating co-ordinate system.

As a first step, consider a vector **A** which is fixed in a co-ordinate system which is rotating with angular frequency ω_c. This is shown pictorial form in Figure 3.7. In time δt, its end point is displaced by an amount δ**A,** so that in terms of the fixed co-ordinate system

$$\delta\mathbf{A} = \left(\omega\delta t\right)\mathbf{A}\sin\theta = \left(\omega_c\times\mathbf{A}\right)\delta t \tag{16}$$

and the velocity of **A** in the fixed system is

$$\lim_{\delta t\to0}\left(\delta\mathbf{A}/\delta t\right) = d\mathbf{A}/dt = \left(\omega_c\times\mathbf{A}\right) \tag{17}$$

If now **A** is not fixed in the rotating system, but is itself moving at a velocity D**A**/D*t*, its velocity in the laboratory co-ordinate system is

$$d\mathbf{A}/dt = D\mathbf{A}/Dt + \left(\omega_c\times\mathbf{A}\right) \tag{18}$$

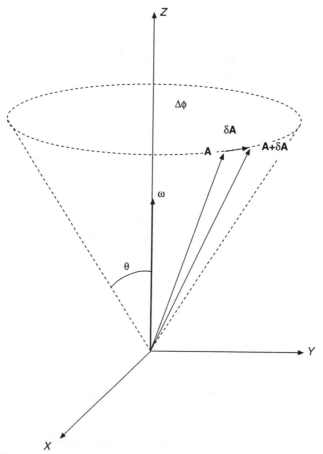

Figure 3.7 Rotating co-ordinate system

Note that the newly introduced notation of the form D**A**/D*t* refers to a separate differentiation operation. Using the form of expression shown in equation 18, we may rewrite the precessional motion as

$$d\mathbf{m}/dt = D\mathbf{m}/Dt + (\boldsymbol{\omega} \times \mathbf{m}) \tag{19}$$

and now substituting this result into equation 12 we obtain

$$D\mathbf{m}/Dt = \gamma\, \mathbf{m} \times \mathbf{B} - (\boldsymbol{\omega} \times \mathbf{m}) \tag{20}$$

$$= \gamma\, \mathbf{m} \times \mathbf{B} + \mathbf{m} \times \boldsymbol{\omega} \tag{21}$$

$$= \gamma\, \mathbf{m} \times \left(\mathbf{B} + \frac{\boldsymbol{\omega}}{\gamma} \right) \tag{22}$$

This expression demonstrates that in a rotating co-ordinate system, the body is subjected to an apparent magnetic field given by ($\mathbf{B}_{app}=\mathbf{B}+\boldsymbol{\omega}/\gamma$), and that the apparent rotational velocity is decreased by the velocity of rotation of the co-ordinate system. We may now remove terms which become constant in the rotating reference frame.

3.5.2. Resonant Motion

We now apply a circularly polarised magnetic field $\mathbf{B}_1$ in a plane normal to the steady field $\mathbf{B}_z$ and view this from within the rotating co-ordinate system. Note that we may decompose a circularly polarised field into two counter-rotating sinusoidal fields of the same frequency. If the additional field $\mathbf{B}_1$ rotates at the same frequency as the new co-ordinate system, the spinning particle experiences an apparent field in the sense of $\mathbf{B}_z$ which is denoted $\mathbf{B}_{app}$. It would be seen to precess (by an observer in the rotating system) about the resultant of $\mathbf{B}_1$ and $\mathbf{B}_{app}$, namely $\mathbf{B}_{res}$. These fields are shown schematically in Figure 3.8.

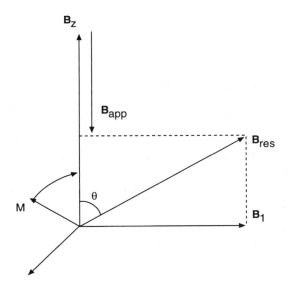

Figure 3.8 Summation of fields in the rotating co-ordinate system.

B_{res} reduces to B_1 when $B_z = B_{app}$. The magnetic moment **m** then rotates around B_1, becoming parallel and antiparallel to B_z. In this condition, the precession frequency

$$\omega_p = -\gamma B_1 = \omega_L \tag{23}$$

has the same frequency as the natural oscillation of precession of the particle's magnetic moment (the Larmor frequency). This is a forced resonance condition, in which the frequency of resonance is proportional to the applied field B_z.

3.5.3. Relaxation Processes

Forcing energy is delivered as a pulse of electromagnetic radiation with energy in the resonant frequency region. Once forced into a resonance condition, the energy acquired by magnetic dipoles requires a time to allow it to be given up to the surrounding material. The resonant effect is then observed by examining the release of that energy to the surrounding material as the nuclear spin returns to alignment along the B_z axis. Firstly we see from the diagram that the magnetisation **M** rotates in the resonant condition about the forcing function B_1. For a field strength of around 10^{-1} T, the precessional rate is in the order of 10^6 rad s^{-1}. This means that it is necessary to administer pulses in the order of 1 µs duration.

We have so far described the resonance phenomenon from the viewpoint of a single spinning particle. We now describe the system in terms of the net magnetisation **M** which is the sum Σm_i over all nuclei in a unit volume.

The first form of decay process to observe is the spin-lattice relaxation time T_1. This is the process in which the stimulated nuclei (normally in our cases protons) release their excess energy to the lattice so that the system returns to a thermodynamic balance. The relaxation process of the magnetisation **M** is described by

$$\frac{d\mathbf{M}}{dt} = \frac{(\mathbf{M}_0 - \mathbf{M})}{T_1} \tag{24}$$

$$\text{or } \mathbf{M}_0 - \mathbf{M} = \mathbf{M}_0 e^{(-t/T_1)}$$

and M_0 the equilibrium magnetisation. This relaxation time is about 2 seconds for water, but values are typically in the range between 10^{-4} and 10^4 s. The relaxation processes use a number of different physical mechanisms by which energy is transferred to the lattice from the resonating nuclei: see Lerski (1985) for a description of various physical models.

In addition to this effect, the spins of neighbouring nuclei may interact. A precessing nucleus produces a local field disturbance $\approx 10^{-4}$ T in its nearest neighbour in water causing a dephasing of protons in $\sim 10^{-4}$ s owing to their frequency differences. The spin-spin interaction time is commonly denoted T_2.

These relaxation processes effectively limit the rate at which an image may be acquired using NMR and its spectral resolution. T_1 means that having stimulated one region, the signal from that area must decay before another area may be stimulated in order to determine its proton population.

3.6. Ultrasound

Sound is the perception of pressure fluctuations travelling through a medium; its waves are transmitted as a series of compressions and rarefactions. There are a number of ways in which this pressure fluctuation can be transmitted which give rise to three classes of wave which are outlined below.

Ultrasound is defined as sound above the range of hearing of the human ear. This is usually taken to be 20 kHz although the appreciation of sound above 16 kHz is exceptional. Figure 3.9 gives an indication of the classification of sound and some natural and manmade phenomena and uses.

up to 20 Hz	Infra sound
117.1 Hz	Middle C
500 Hz	Underwater Navigation
1.77 kHz	Upper Soprano
16 kHz	Upper Limit of Normal Hearing
20 kHz	Ultrasound
30 kHz	Early Submarine Detection
70 kHz	Upper Limit of Bats
≥70 kHz	Sonar
500 kHz	Lower limit of Non Destructive Testing (NDT)
500 kHz–12 MHz	Medical
	Imaging up to 12 MHz
	Doppler 2,4,6,8 MHz
12 MHz–100 MHz	Scanning Acoustic Microscope (SAM)

Figure 3.9 The sonic spectrum

3.6.1. Longitudinal or Pressure Waves

In a Longitudinal wave the particles of the transmission medium move with respect to their rest position. The particle movement causes a series of compressions and rarefactions. The wave front travels in the same direction as the particle motion. The particle movement and subsequent compressions cause corresponding changes in the local density and optical refractive index of the material of the medium.

3.6.2. Shear or Transverse Waves

In shear waves, the wave front moves at right angles to the particle motion. Shear waves are often produced when a longitudinal wave meets a boundary at an oblique angle.

3.6.3. Surface, Rayleigh or Lamb Waves

Rayleigh or Lamb waves occur at the surface of materials and only penetrate a few wave-lengths deep. These waves occur only in solids. Some semiconductor filters have been developed which rely on the properties of surface waves travelling in crystalline materials.

For medical applications we need only consider longitudinal waves as both Imaging and Doppler techniques rely on the propagation of longitudinal waves. Shear waves can propagate in fluids: however, they are not intentionally produced.

3.7. Physics of Ultrasound

3.7.1. Velocity of the Propagating Wave

The velocity (c) of a longitudinal wave travelling through a fluid medium is given by the ratio of its bulk modulus to its density.

$$c = \sqrt{\frac{K}{\rho}}$$

(25)

where

K = bulk modulus

ρ = density

3.7.2. Characteristic Acoustic Impedance

The relationship between particle pressure and the particle velocity is analogous to Ohm's law. Pressure and velocity correspond to voltage and current respectively. The acoustic impedance is therefore a quantity analogous to impedance in electrical circuits. It is related to particle pressure and velocity by the following equation:

$$p = Zv$$

(26)

where

p= particle pressure

v= particle velocity

Z= acoustic impedance

Acoustic impedance can be expressed as a complex quantity in the manner of electrical impedance. However for most practical medical applications it can be considered in a simple form. The characteristic acoustic impedance of a material is the product of the density and the speed of sound in the medium:

$$Z = \rho c$$

(27)

where ρ = density in kg m^{-3}.

Hence, materials with high densities have high acoustic impedances. For instance steel has a higher acoustic impedance than perspex. The following table shows materials with similar and dissimilar acoustic impedances.

Similar Z

Iron – Steel

Water – Oil

Fat – Muscle

Dissimilar Z

Water – Air

Steel – Fat

The dimensions of the acoustic impedance are kg m^{-2} s^{-1}. Most materials found in the human body or used in transducers have acoustic impedances of the order of 10^6 kg m^{-2} s^{-1}; therefore, the commonly expressed unit of acoustic impedance is the Rayle. One Rayle is 1×10^6 kg m^{-2} s^{-1} .

The acoustic impedance of a number of materials is presented in Figure 3.10.

Material	Velocity ms^{-1}	Density kgm^{-3}	Acoustic Impedance 10^6kgm^{-2}s^{-1}
Steel	7900	5800	45.8
Bone	3760	1990	7.48
Skin	1537	1100	1.69
Muscle	1580	1041	1.64
Fat	1476	928	1.36
Blood	1584	1060	1.68
Water	993	1527	1.52
Air	330	1.2	0.0004

Figure 3.10 Table of acoustic impedance values (see Wells 1977, Duck 1990)

3.7.3. Acoustic Intensity

Consider a particle vibrating with Simple Harmonic Motion (SHM) in a lossless medium. The total energy of the particle (e_{total}) is the sum of its potential and kinetic energies. If the medium is lossless the total energy is constant. The total energy of the particle when at zero displacement from its resting position is given by its kinetic energy:

$$e_{total} = \frac{1}{2}m v_0^2 \tag{28}$$

where v_0 = velocity when at zero displacement

m = particle mass

The total mass of particles contained within unit volume is given by the density of the medium (ρ). Therefore the total energy of the particles in unit volume is given by

$$E_{TUV} = \frac{1}{2}\rho v_0^2 \tag{29}$$

The intensity (I) of a wave can be defined as the energy passing through unit area in unit time.

The wave velocity is the rate at which this particle energy passes through the medium. Therefore in unit time a unit area will travel a distance of c metres, defining a volume c. As the total energy per unit volume is E_{TUV}, the energy passing through unit area in unit time will be given by

$$I = c\ E_{TUV} = \frac{1}{2}c\rho v_0^2 \tag{30}$$

The intensity can also be expressed in terms of pressure.

$$I = \frac{1}{2}c\rho\,v_0^{\,2} \equiv \frac{1}{2}Z\,v_0^{\,2} \equiv \frac{Z^2\,v_0^{\,2}}{2Z} \equiv \frac{p^2}{2Z} \tag{31}$$

This equation's dimensions are: $I = $ m s^{-1} × kg m^{-3} × m^2 s^{-2} = kg s^{-3}.

The units of intensity are watts per square metre, which is equivalent to kg s^{-3}.

3.7.4. Reflection

If a longitudinal wave travelling through a medium meets an interface with a different medium, reflection or transmission of the wave will occur. The laws of geometric reflection can be applied as long as the wavelength of the ultrasound is small compared to the dimensions of the interface. If this is so the reflection is said to be 'specula'. However, if this condition does not apply then scattering occurs. This will be considered in section 3.7.7.

Consider a wave travelling through a medium and impinging upon an interface at an angle θ_i (Figure 3.11), a portion of the wave will be reflected at an angle θ_r equal to the angle of incidence. Some of the wave is transmitted at an angle θ_t given by Snell's law.

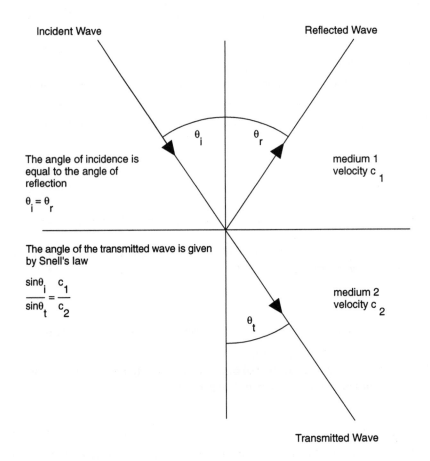

Figure 3.11 Snell's law

$$\frac{\sin\theta_i}{\sin\theta_t} = \frac{c_1}{c_2} \tag{32}$$

where c_1 and c_2 are the velocities of the wave in media 1 and 2 respectively. The subscripts i, t, r refer to the incident, transmitted and reflected waves respectively.

For a particular interface, as the angle of incidence increases, the angle of transmission also increases until the point of total internal reflection is reached. Total internal reflection occurs when the angle of the transmitted wave is equal to $\pi/2$. Therefore from equation (32) the incident angle for total reflection to occur is given by:

$$\theta_i = \sin^{-1}\frac{c_1}{c_2} \qquad \text{as} \quad \sin\pi/2 = 1 \tag{33}$$

if $\qquad c_2 > c_1$

3.7.4.1. Pressure Relationship

The particle pressure at an interface must be continuous. Therefore the sum of the particle pressure on one side is equal to that on the other or

$$p_i + p_r = p_t \tag{34}$$

Consider a wave with particle velocity v_i impinging upon an interface at an angle θ_i. The velocity either side of the interface is also continuous and therefore.

$$v_i \cos\theta_i - v_r \cos\theta_r = v_t \cos\theta_t \tag{35}$$

As the particle velocity is a vector, the reflected velocity is negative (in the opposite direction) with respect to the incident wave.

Recalling equation (26) equations (34) and (35) can now be combined

$$p_i/Z_1 \cos\theta_i - p_r/Z_1 \cos\theta_r = p_t/Z_2 \cos\theta_t \tag{36}$$

p_r/p_i is known as the pressure reflectivity and p_t/p_i is known as the pressure transmittivity. Equation (36) can be solved to yield:

$$\frac{p_r}{p_i} = \frac{Z_2\cos\theta_i - Z_1\cos\theta_t}{Z_2\cos\theta_i + Z_1\cos\theta_t} \tag{37}$$

and

$$\frac{p_t}{p_i} = \frac{2Z_2\cos\theta_i}{Z_2\cos\theta_i + Z_1\cos\theta_t} \tag{38}$$

These equations are often shortened by assuming the incidence to be normal so all the cosine terms are 1. Therefore equations (37) and (38) reduce to:

$$\frac{p_r}{p_i} = \frac{Z_2 - Z_1}{Z_2 + Z_1} \quad \text{and} \quad \frac{p_t}{p_i} = \frac{2Z_2}{Z_2 + Z_1} \tag{39}$$

There will therefore be no reflection at an interface between two materials if their acoustic impedances are equal.

Consider an ultrasound wave travelling from medium 1 to medium 2 with acoustic impedances Z_1 and Z_2 respectively. If $Z_1 > Z_2$ the reflected wave will be π radians out of phase with
the incident wave. However, if $Z_1 < Z_2$ the reflected wave will be in phase with the incident
wave.

3.7.4.2. Intensity Relationship

The preceding equations define the transmission of a pressure wave across a boundary. By
following the derivation for obtaining pressure expressions we may arrive at equations which
define the intensity of waves at a boundary. Recall equation (31) which may be substituted
into equations (37) and (38) to describe the wave intensity.

$$\frac{I_r}{I_i} = \left(\frac{Z_2 \cos\theta_i - Z_1 \cos\theta_t}{Z_2 \cos\theta_i + Z_1 \cos\theta_t} \right)^2 \quad \text{and} \quad \frac{I_t}{I_i} = \frac{4 Z_1 Z_2 \cos\theta_i}{\left(Z_2 \cos\theta_i + Z_1 \cos\theta_t \right)^2} \tag{40}$$

where I_r/I_i is known as the intensity reflectivity and I_t/I_i is known as the intensity
transmittivity. These equations are often simplified by assuming normal incidence so equating all the cosine terms to 1. Therefore equation (40) becomes:

$$\frac{I_r}{I_i} = \left(\frac{Z_2 - Z_1}{Z_2 + Z_1} \right)^2 \quad \text{and} \quad \frac{I_t}{I_i} = \frac{4 Z_1 Z_2}{\left(Z_2 + Z_1 \right)^2} \tag{41}$$

Hence, the degree of transmission or reflection of the pressure or the intensity of an acoustic
wave incident on a boundary between two materials is related to their acoustic impedances.
Recalling the table of material pairs with similar and dissimilar acoustic impedances, clearly
there will be minimal transmission and almost total reflection between the dissimilar materials. Conversely negligible reflection and almost total transmission occurs between similar
materials.

Reflections from soft tissue

   Kidney / Muscle = 0.03;   Soft Tissue / Bone = 0.65;   Tissue Air Coupling = 0.999

3.7.4.3. Transmission Through Thin Layers

The preceding analysis determined equations relating the intensity of a wave incident on an
interface to the acoustic impedance of the two materials. The transmission of ultrasound
through a thin layer is given by the following equation. It is a special case and will be
considered as it has important implications for transducer design and practical application of
ultrasound in medicine (Hill 1986).

$$T = \frac{4 Z_3 Z_1}{\left(Z_1 + Z_3 \right)^2 \cos^2 2\pi \frac{t_2}{\lambda_2} + \left(Z_2 + \frac{Z_1 Z_3}{Z_2} \right)^2 \sin^2 2\pi \frac{t_2}{\lambda_2}} \tag{42}$$

Where T is the transmission and t_2 is the thickness of the thin layer with impedance Z_2
between media Z_1 and Z_3.

There are three situations when this equation can be simplified.

1.   If $Z_1 \gg Z_2$ and $Z_3 \gg Z_2$ then the right hand side of the denominator will be large and

therefore the transmission of ultrasound through the thin layer will be negligible. This situation occurs when there is a layer of air trapped between an ultrasound transducer and a patient.

2.   If $\cos^2 2\pi \frac{t_2}{\lambda_2} = 1$ i.e. when $t_2 = n\lambda_2$ where $n = 1,2,3,4,5,6,\ldots$ then

$$T = \frac{4Z_1 Z_3}{(Z_1 + Z_3)^2}$$

In this instance the thickness of the thin layer is chosen such that transmission through it is independent of its acoustic properties. This is known as a half wave matching layer.

3.   If $\sin^2 2\pi \frac{t_2}{\lambda_2} = 1$ i.e. when $t_2 = (2n-1)\frac{\lambda_2}{4}$ where $n = 1,2,3,4,5,6\ldots$ then

$$T = \frac{4Z_1 Z_3}{\left(Z_2 + \dfrac{Z_1 Z_3}{Z_2}\right)^2}$$

If the impedance of the second material can be chosen such that it is equal to $Z_2 = \sqrt{(Z_1 Z_3)}$ then the transmission through the layer can be total. This situation is known as a quarter wave matching layer.

Both quarter and half wave matching layers are used in ultrasonics (section 3.9.3); however, the properties of these layers depend on the wavelength in the second medium and therefore as the wavelength changes with frequency they are frequency specific.

3.7.5. Attenuation

So far we have referred to the conducting medium for ultrasonic propagation as lossless. However, in all practical situations the intensity of a wave diminishes with its passage. The reduction in the intensity or pressure of a wave passing through a medium in the x direction is referred to as the attenuation of the medium. The reduction in the wave can be attributed to a number of effects: namely reflection, wave mode conversion (longitudinal to shear), beam spreading, scattering and absorption. Attenuation varies with frequency as both scattering and absorption are frequency dependent.

The attenuation of a medium is expressed in terms of dB cm^{-1} at a particular frequency. Attenuation can be determined for the pressure or intensity of a wave. The intensity attenuation coefficient is given by

$$A_I = \frac{10}{x} \log \frac{I_1}{I_2} \tag{43}$$

and the pressure attenuation coefficient by

$$A_P = \frac{20}{x} \log \frac{P_1}{P_2} \tag{44}$$

In each case x is the displacement between the points 1 and 2 where intensity and pressure I_1, P_1 and I_2, P_2 were measured.

3.7.6. Absorption

An ultrasonic wave travelling through a medium is absorbed when wave energy is dissipated as heat. Absorption occurs when the pressure and density changes within the medium caused by the travelling wave become out of phase. When this happens wave energy is lost to the medium. The fluctuations become out of phase with the density changes as the stress with the medium causes the flow of energy to other forms. In section 3.7.3 we derived an expression for the intensity of a wave travelling through a lossless medium by considering the energy of a particle to be composed entirely of potential and kinetic energy.

In a real medium, the total wave energy is shared between a number of forms which include molecular vibration and structural energy. During the compression cycle of the longitudinal wave, mechanical potential energy is transferred to other forms. During the rarefaction of the medium the energy transfer reverses and the energy is returned to the wave. The energy transfer is referred to as a relaxation process.

The relaxation process takes a finite amount of time, known as the relaxation time (the inverse of which is known as the relaxation frequency). If the wave is at low frequency then the energy transfer can be completed. However, as the frequency increases, the energy transfer becomes out of phase with the wave, energy is lost and absorption occurs. The absorption increases with frequency reaching a maximum at the relaxation frequency. At frequencies above the relaxation frequency the absorption decreases as there is insufficient time for the initial energy transfer to take place.

Figure 3.12a shows the variation of absorption with frequency for a single relaxation process. If one considers two relaxation processes with different relaxation frequencies, one would find that, generally, the higher frequency process would cause greater absorption. This situation is depicted in Figure 3.12b.

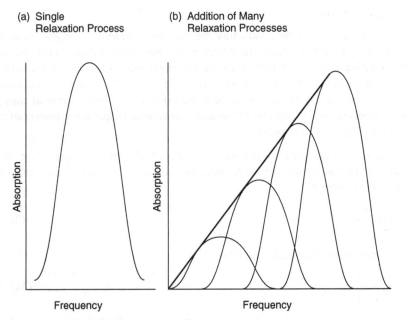

(a) Single
     Relaxation Process

(b) Addition of Many
     Relaxation Processes

Figure 3.12 Relaxation process

In biological materials there is a large number of different relaxation processes, each of which has a characteristic differing relaxation frequency. Therefore, the absorption characteristic of tissue increases approximately linearly with frequency and is attributable to the summation of absorption from a large number of relaxation processes.

3.7.7. Scattering

If a wave with wavelength λ impinges upon a boundary whose dimensions are large compared to the wavelength, then specular reflection will occur. However, if the obstacle is smaller than the wavelength or of comparable size the laws of geometric reflection will not apply. In this instance, the wave is said to scattered using one of two different processes, Rayleigh and Stochastic.

1. The Rayleigh region is when the dimensions of the scattering object are very much less than the wavelength of the incident ultrasound. In the Rayleigh region incident ultrasound is scattered equally in all directions. The relationship determining the degree of scattering is the same as that derived for light. See, for example, Longhurst (1967).

$$\text{Scattering} \propto \left(\frac{2\pi}{\lambda}\right)^4 \propto f^4$$

2. If the dimensions of the scatterer are similar to the wavelength of the incident ultrasound then the scattering is stochastic. In this region there is a square law relationship between the degree of scattering and frequency.

The ratio of the incident ultrasonic intensity to the power scattered at a particular angle is known as the scattering cross section. If S_I is the power of the scattered ultrasound and I_I is the intensity of the incident ultrasound then α, the scattering cross section, is given by

$$\alpha = \frac{S_I}{I_I}$$

In Doppler blood flow detection and in medical imaging the majority of the detected signal originates from scattered ultrasound. Therefore the variation of scattering with angle is of importance. The ratio of the intensity of the ultrasound scattered at a particular angle to the intensity of the incident ultrasound is the differential scattering cross section (the scattering cross section at a particular angle). Of most importance in medical imaging and Doppler blood flow studies is the scattering cross section at 180°, which corresponds to ultrasound transmitted directly back to the source as this determines the signal detected by the system.

3.7.8. Attenuation in Biological Tissues

The attenuation in biological materials has been measured both in vivo and in vitro. Tests are conducted at a given temperature, pressure and frequency. The standard values determined may find some clinical importance: for example, attenuation in tumour tissue is different from attenuation in breast tissue. However, attenuation by tissue is not at present used routinely in clinical situations. The attenuation of various tissues is represented in Figure 3.13. These values are important when designing any ultrasound system as they determine the strength of the echoes received from a certain depth in either ultrasonic imaging or Doppler studies.

Material	Attenuation dB cm^{-1}
Skin	3.5 ± 1.2
Bone	13
Muscle	2.8
Fat	1.8 ± 0.1
Blood	0.21

Figure 3.13 Table of attenuation values (Duck 1990)

3.8. The Doppler Effect

3.8.1. Introduction

The Doppler effect was first derived in 1845 by the German physicist C.J. Doppler (1803–1853). He noted that there was a change in the detected frequency when a source of sound moved relative to an observer.

The Doppler effect will have been noticed by readers as the world we live in is full of examples of the slight change in the sound detected from a moving object. For example, when an ambulance with a siren or a motor bike passes, the note we hear is affected by the velocity of the source.

The sounds we hear are characterised by their frequencies. When a sound is emitted from a moving source the apparent frequency a stationary observer detects is affected. The apparent frequency will increase if the velocity of an emitter is positive, towards the detector, conversely the frequency will be lowered if the velocity is negative (the sign of the frequency shift is therefore dependent on the sign of the velocity). This is why the effect is most noticeable when the source passes us, as the velocity becomes negative and the Doppler shift suddenly changes from being positive to negative. The magnitude of the Doppler effect depends on the magnitude of the velocity.

The Doppler effect has been used for many years for military and commercial Radar allowing the velocity and the position of an aeroplane to be determined. In medicine, Doppler techniques have been substantially developed for blood flow studies enabling determination of blood flow velocity, detection of turbulence associated with pathological disturbances and the detection of foetal heart beats.

3.8.2. Derivation Of Doppler Equations

3.8.2.1. Stationary Detector Moving Source

Figure 3.14 is a diagrammatic representation of the effect of the moving source. If the velocity is away from the detector then the apparent wavelength is increased. Conversely movement towards the detector shortens the apparent wavelength and increases the frequency.

Think of an object emitting sound moving directly away from an observer and at constant velocity. Then the apparent wavelength detected by the observer will be elongated by the distance that the source moves while that wave is being emitted.

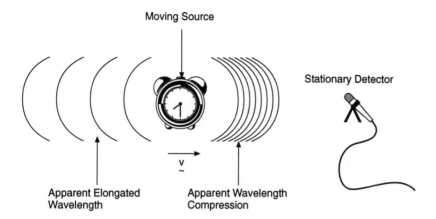

Moving Source

Stationary Detector

Apparent Elongated
Wavelength

Apparent Wavelength
Compression

Figure 3.14 Moving source

the velocity of sound in the medium is c ms^{-1}
the velocity of the source is v ms^{-1}
the frequency emitted from the source is f Hz
the wavelength of the emitted wave is λ metres
the apparent wavelength of the detected wave is λ_a metres

$$c = f_s \lambda_s \tag{45}$$

$$f_s = \frac{c}{\lambda_s} \tag{46}$$

The apparent wavelength is the distance travelled by the wave front in time Δt divided by the number of oscillations in time Δt.

$$\lambda_a = \frac{\text{displacement in } \Delta t}{\text{number of oscilations in } \Delta t} \tag{47}$$

$$\lambda_a = \frac{(c + v)\,\Delta t}{f_s\,\Delta t} \tag{48}$$

$$f_a = \frac{c}{\lambda_a} \tag{49}$$

So

$$f_a = \frac{c f_s\,\Delta t}{(c + v)\,\Delta t} \tag{50}$$

Cancel the factor Δt

$$f_a = f_s \left(\frac{c}{(c + v)} \right) \tag{51}$$

Divide by c

$$f_a = f_s \left(\frac{1}{(1 + \frac{v}{c})} \right)$$

(52)

This is the Doppler equation for a moving source, the sign of the denominator is positive for movement away from the detector and negative for movement towards the detector.

3.8.2.2. Special case for $v \ll c$

If the velocity of the source v is small compared with the velocity of the wave in the medium, then equation (52) can be simplified by using a series expansion.

The series expansion of

$$y = \left(\frac{1}{(1 + \frac{v}{c})} \right)$$

(53)

is

$$y = 1 + \left(\frac{v}{c}\right) + \left(\frac{v}{c}\right)^2 + \left(\frac{v}{c}\right)^3 + \left(\frac{v}{c}\right)^4 \ldots$$

(54)

But if $v \ll c$ then equation (54) approximates to

$$y \approx \left(1 + \frac{v}{c}\right)$$

(55)

Therefore equation (52) can be rewritten as

$$f_a \approx f_s \left(1 + \frac{v}{c}\right)$$

(56)

The Doppler frequency is often expressed in terms of the Doppler shift f_d where

$$f_d = f_a - f_s$$

(57)

Combining (56) and (57)

$$f_d = \frac{f_s v}{c}$$

(58)

In Doppler analysis of blood flow $v \ll c$ as the velocity of sound is approximately 1500 ms^{-1} in most soft tissues and the blood flow velocity is in the range 0 to 5 ms^{-1}.

The Doppler shift is therefore directly proportional to the source frequency, the source velocity and inversely proportional to the speed of sound in the medium of interest.

3.8.2.3. Moving Detector and Stationary Source

Consider a stationary source emitting sound waves and a detector moving with velocity v in a straight line towards the source as depicted in Figure 3.15. The wavelength in this situation stays constant and the apparent velocity increases causing the detector to cross a greater number of wave crests leading to a consequently increased frequency. Hence the frequency increases by the extra number of waves received in time Δt.

$$f_a = f_s + \frac{\text{extra number waves received in time } \Delta t}{\Delta t} \tag{59}$$

$$f_a = f_s + \frac{v \, \Delta t \, / \lambda_s}{\Delta t} \tag{60}$$

$$f_a = f_s + \frac{v}{\lambda_s} \tag{61}$$

But

$$\lambda_s = \frac{c}{f_s} \tag{62}$$

$$f_a = f_s + \frac{v f_s}{c} \tag{63}$$

$$f_a = f_s \left(1 + \left(\frac{v}{c} \right) \right) \tag{64}$$

Combining (57) and (63)

$$f_d = \frac{f_s \, v}{c} \tag{65}$$

This is the same result as for a moving source.

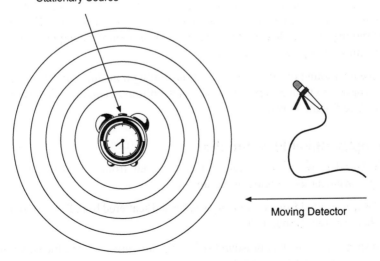

Stationary Source

Moving Detector

Figure 3.15 Moving detector

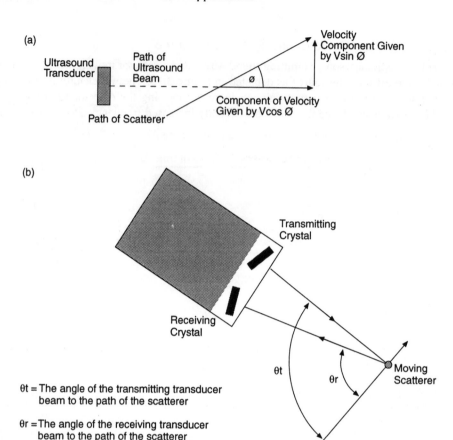

Figure 3.16 Velocity vector

When considering Doppler shifts caused by moving interfaces, the vector component of the velocity has to be considered. Therefore, only the component of the scattering object's velocity acting along the axis of the ultrasound wave is considered (see Figure 3.16a). If a scatterer is moving at a angle ϕ to the ultrasound beam then the component of the velocity acting along the axis is given by $v\cos\phi$.

Normally, the ultrasound probe consists of a transmitting transducer and a separate receiving transducer. These two transducers necessarily make different angles to the velocity vector of the scatterer (see Figure 3.16b).

3.8.3. Doppler Blood Flow Studies

In Doppler blood flow studies ultrasound is scattered from a moving object back to a transmitting transducer as depicted in Figure 3.17.

Ultrasound is transmitted from a stationary source and scattered back from a moving scatterer. Therefore there are two Doppler shifts:

- The first shift occurs as the ultrasound strikes the moving scatterer, the situation is that of a stationary source and a moving detector.

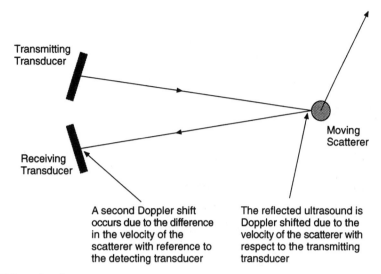

Transmitting
Transducer

Receiving
Transducer

Moving
Scatterer

A second Doppler shift
occurs due to the difference
in the velocity of the
scatterer with reference to
the detecting transducer

The reflected ultrasound is
Doppler shifted due to the
velocity of the scatterer with
respect to the transmitting
transducer

Figure 3.17 Doppler flow measurement

• The second frequency shift occurs as the ultrasound is scattered back, this situation is that of a stationary detector and a moving source.

This can perhaps be appreciated if you consider the radar detection of an aeroplane's velocity. If a radar signal is transmitted from the ground, the frequency detected by the plane's radar operator will be shifted due to the plane's velocity (the first Doppler shift). The radar signal reflected back to the ground will neither be that originally transmitted nor that detected on the plane but rather a further shifted signal (the second Doppler shift).

Consider the situation where an object moves towards the ultrasound probe. The first Doppler shift will be given by equation (64).

$$f_a = f_s\left(1+\frac{v_s}{c}\cos\theta_t\right)$$

(66)

This can now be substituted into equation (56) as the source frequency

$$f_a = f_s\left(1+\left(\frac{v_s}{c}\cos\theta_t\right)\right)\left(1+\left(\frac{v_s}{c}\cos\theta_r\right)\right)$$

(67)

Multiplying and equating the squared term to zero, as $v \ll c$ leaves:

$$f_a = f_s\left(1+\frac{v_s}{c}\cos\theta_t+\frac{v_s}{c}\cos\theta_r\right)$$

(68)

Recalling equation (57)

$$f_d = +\frac{f_s v_s\left(\cos\theta_t+\cos\theta_r\right)}{c}$$

(69)

$$f_d = -\frac{2f_s v_s}{c}\left(\cos\frac{\theta_t + \theta_r}{2}\cos\frac{\theta_t - \theta_r}{2}\right) \tag{70}$$

The angles θ and ϕ are defined by Hill (1986) (see Figure 3.18)

$$f_d = +\frac{2f_s v_s}{c}\cos\theta\cos\frac{\phi}{2} \tag{71}$$

But $\cos\left(\frac{\phi}{2}\right) \approx 1$

and therefore equation (71) can be rewritten as

$$f_d = +\frac{2f_s v_s}{c}\cos\theta \tag{72}$$

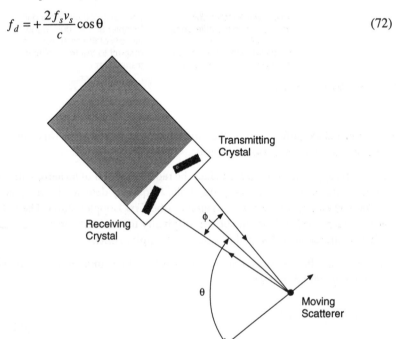

Figure 3.18 Doppler scattering angles

3.9. Generation and Detection of Ultrasound

Ultrasound is defined as acoustic vibration above the range of human hearing. As the frequency lies between 20 kHz and 20 MHz there is a variety of methods of generation and detection. At low frequencies ultrasound may be generated by specialised loud speakers and sirens. At higher frequencies however, these mechanisms become difficult, and piezo electric and magneto strictive transducers are used.

In medical application with frequencies between 1 and 12 MHz piezo electric transducers are used for both generation and detection of ultrasound.

3.9.1. Piezo Electric Materials

When a piezo electric material is stressed a potential difference is produced. Conversely when a potential is applied a stress is produced within the material. These two phenomena are exploited respectively in the detection and generation of ultrasound. Figure 3.19 demonstrates how a piezo electric material may be used to generate and detect ultrasound.

Piezo electric properties are due to the chemical morphology of the material. In a piezo electric material the constituent molecules are electrically polarised i.e. one end of the molecule is electrically negative and the other end positive. The polarisation is due to the unequal sharing of electrons within the molecule. For example when fluorine combines with other elements to form a molecule the fluorine has a strong attraction for electrons and

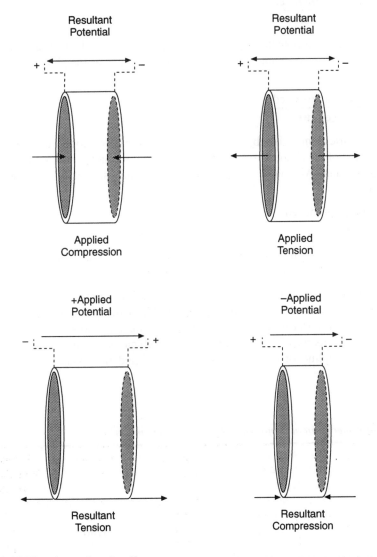

Figure 3.19 The piezo electric effect

therefore is electro negative with respect to the rest of the molecule. Electrically polarised molecules are referred to as dipoles In naturally occurring materials the electrical dipoles are normally randomly arranged and the material therefore exhibits weak piezo electric properties.

3.9.1.1. Electro Mechanical Properties of Piezo Electric Elements

The characteristics of piezo electric materials are classified by a number of coefficients.

The electromechanical coupling factor (k^2) is the ratio of the mechanical energy converted to electrical energy to the total mechanical energy input to the material.

Conversely the factor may be defined as the ratio of electrical energy converted to mechanical energy to the total electrical energy input to the material.

$$k^2 = \frac{\text{energy converted}}{\text{total energy}} \tag{73}$$

The ratio of the electrical potential developed by a piezo electric element in response to an applied mechanical stress is denoted by the coefficient g known as the piezo electric voltage constant.

$$g = \frac{\text{potential produced}}{\text{applied stress}} \tag{74}$$

When a piezo electric element is subjected to a force a charge is developed, the coefficient d is defined as the ratio of the charge produced to the applied force and is referred to as the piezo electric charge constant. Alternatively d may be defined as the resultant deformation due to an applied potential.

$$d = \frac{\text{charge produced}}{\text{applied force}} \tag{75}$$

The value of g should be high for a piezo electric element used in a receiving mode. Generally it is desirable for transducer materials to have high values of k^2, d and g.

A transducer element exhibits piezo electric behaviour in three dimensions. The coefficients k^2, d and g can be expressed in terms of the relationship between quantities applied in the same or different planes. For instance k^2 is expressed for thickness mode vibrations as k^2_T or k^2_{33} where the subscripts refer to the 3rd plane of the element or the thickness. Alternatively k^2_{31} refers to the coupling of energy between the third and first plane of the transducer element.

Table 3.1 Electro mechanical properties of piezo electric materials

Materials	$k2$	d VmN^{-1}	g mV^{-1}	Velocity ms^{-1}	Acoustic Impedance Rayle
PZT	0.7	289	2.61	4000	30
Quartz	0.1	2.31	5.78	5740	15
PVDF	0.19				15

The constants d, g and k^2 vary with temperature. A material's piezo electric properties disappear at the Curie Point, and are not constant with time. In addition a piezo electric element may become depolarised by high stresses and high alternating fields.

3.9.1.2. Types of Piezo Electric Materials

There are naturally occurring piezo electric materials such as quartz which display piezo electric properties when cut along a particular plane to obtain domain alignment. Quartz has been used for many years as a piezo electric material in transducers. Some biological materials, such as bone, are weakly piezo electric. However, the majority of piezo electric materials used in ultrasonic transducers are manmade. There are three types of manmade piezo electric material: crystalline, ceramic and polymer.

Crystalline materials such lithium niobate are grown in solution and are, therefore, difficult to produce and shape by machining. Ceramic materials such as lead zirconium titanate (PZT) and Polymer materials such as polyvinyl diflouride (PVDF) are heated close to a material dependent temperature known as the Curie Point or Temperature. When heated above this temperature the piezo electric domains disappear, reappearing when the material temperature falls. If the material temperature is close to the Curie Point the domains can move and therefore a potential may be applied to attain domain alignment as the material is cooled. Polymer materials are stretched to increase their piezo electric properties.

Ceramic materials have high piezo electric coefficients when compared to polymer materials. However, the acoustic impedance of ceramic materials (approximately 30 Rayle) is an order of magnitude greater than that of soft body tissues (fat approximately 1.5 Rayle). Ceramic materials produce high mechanical and electrical responses when used as either generators or detectors of ultrasound. However they suffer from poor transmission of ultrasound through the tissue-transducer interface. Polymer materials have an acoustic impedance well matched to the acoustic impedance of body tissues but have low piezo electric coupling factors (k^2).

To produce a material with good piezo electric properties and low acoustic impedance some further transducer materials have been developed which are called copolymers. These consist of powdered piezo electric ceramic within a polymer material.

3.9.1.3. Transducer Properties

In medical applications transducer elements are formed from discs of the piezo electric material (Figure 3.20). When a piezo electric element is excited it can vibrate in three dimensions. The required vibration for an ultrasonic transducer in a medical application is in thickness mode, that is vibrating along the x axis in Figure 3.20. The disc is supported around its rim and polarised along the x axis.

When a transducer is excited ultrasound is emitted from both faces of the disc. Part of the energy arriving at the front face is reflected back into the transducer element. This reflected signal is itself partially reflected by the rear face and returned in the original direction of the wave. Some of the energy arriving at the front face will therefore have travelled a distance equal to twice the disc thickness. If this extra trip distance is equal to one wavelength, constructive interference takes place.

The frequency which results from that wavelength is the natural or resonant frequency of the disc. The disc may resonate at odd multiples of this frequency giving rise to harmonics. There

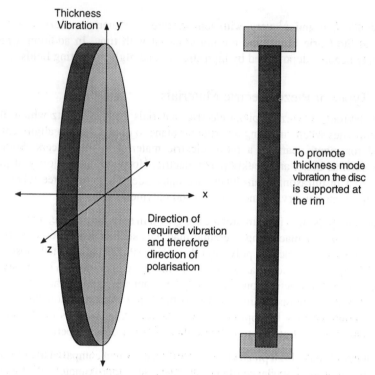

Thickness
Vibration

y

x

z

Direction of
required vibration
and therefore
direction of
polarisation

To promote
thickness mode
vibration the disc
is supported at
the rim

Figure 3.20 Mounting crystals

is therefore a number of resonant frequencies of a transducer disc. However, the strength of the resonance decreases with increasing frequency. The transducer element exhibits resonant behaviour both when transmitting and receiving ultrasound.

The quality of resonance of any system can be described in terms of the width of its half energy amplitude (–3 dB) (see Figure 3.21). This factor is termed the Quality or Q of the resonance and is defined as:

$$Q = \frac{f_{peak}}{f_{upper} - f_{lower}} \qquad (76)$$

A high Q system is by definition of narrow bandwidth.

The resonance of a transducer element is restricted by losses in the material. Crystalline and ceramic materials have low internal losses and therefore resonate strongly. In contrast polymer materials have high losses and a weak resonance and, therefore, exhibit a broad frequency response with low Q.

3.9.2. Transducers' Characteristics

Transducer characteristics can be damped (to reduce their Q) by backing the disc with an energy absorbing material with an acoustic impedance similar to that of the transducer element. In practice the Q of a transducer can be reduced most easily by backing the element with the same material as the active element but in an un-polarised form, or with tungsten

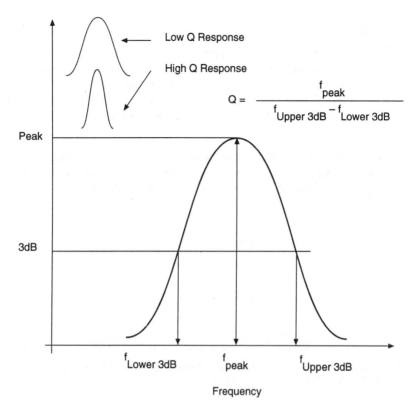

Figure 3.21 Q of resonance

loaded epoxy resin. The energy transmitted through the rear surface of the transducer is then dissipated in the backing. High Q transducers are constructed with air backing which deliberately provides poor coupling.

3.9.2.1. Transducer Fields

The acoustic field produced by a transducer depends on the designed frequency, the active element diameter and shape. The calculation of a transducer's acoustic field is achieved by modelling the surface of the piezo electric crystal as a series of infinitely small piston generators. The field pattern is then calculated by determining the interference between the waves generated by each piston. This method of calculating the transducer field is referred to as Huygen's Method.

The acoustic field produced by a transducer is complicated. Two forms of approximation are used to provide simplified working models. Close to the transducer, the Fresnel approximation provides the better estimates of the field, whereas at greater distances, the Fraunhofer approximation is employed. The intensity calculated using the Fresnel approximation in the vicinity of a transducer of diameter D is an oscillatory pattern over a region of approximately the transducer's dimensions. At depths of $D^2/4\lambda$ the oscillatory pattern gives way to a region of uniformly divergent intensity. At a distance of x_{near} given by

$$x_{near} = D^2 / \lambda \qquad\qquad (77)$$

the half power width of the beam equals the transducer diameter. Whilst the Fresnel model is valid at all depths, the analysis required to obtain it is more complex, so at greater depths the Fraunhofer approximation is appropriate. In this case the width of the beam is defined by $x\lambda/D$ (where x is the distance from the transducer face), which effectively limits the lateral resolution of the transducer. An analytic treatment of these models is provided by Macovski (1983).

Figure 3.22 shows the characteristic field shape produced by a disc transducer when transmitting continuously. The field can be considered as consisting of two regions, the near field and the far field. Figure 3.22 shows these two regions. The limit of the near field can be defined as

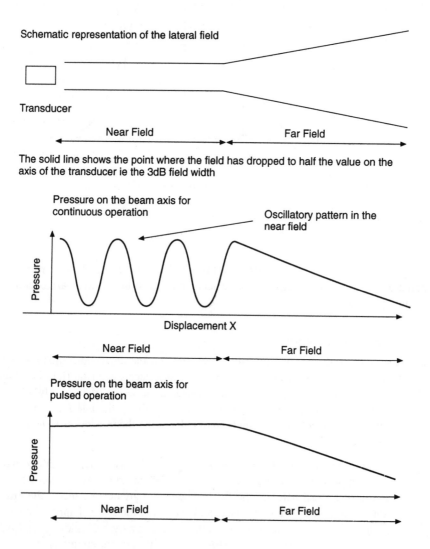

Figure 3.22 Field shapes

the distance from the transducer face when the 3 dB beam width is equal to the diameter of the transducer. The pulsed field distribution is represented schematically in Figure 3.22. It differs from the continuous field as the pressure distribution is approximately uniform in the near field.

In medical applications transducers are designed to perform in the near field region which is up to the point where the field rapidly diverges. Therefore, the maximum depth of scan is effectively controlled by the degradation of the lateral resolution which is related to the disc diameter and operating frequency.

3.9.3. Basic Transducer Design

The basic design of an ultrasonic transducer is shown in Figure 3.23.

1. The choice of piezo electric material was described in section 3.9.1.

2. The electrodes which connect to the transducer are thin layers of either gold or aluminium. The layers are deposited by evaporating metallic wire in a vacuum. The piezo electric material is masked to ensure that the desired regions only are coated with the metal. The thickness of the layer is minimised to reduce interference with the emitted ultrasound.

3. The choice of backing material is dependent on the required use. For Doppler applications a mismatched backing allows the disc to resonate at its natural frequency: this maximises the output power at this specific frequency. In this instance almost no ultrasound is transmitted through the backing so the majority of energy produced is transmitted usefully through the front face of the transducer. Doppler signals have a narrow bandwidth compared to the response of the transducers at high frequencies and so the transducer design can utilise a high Q resonant response.

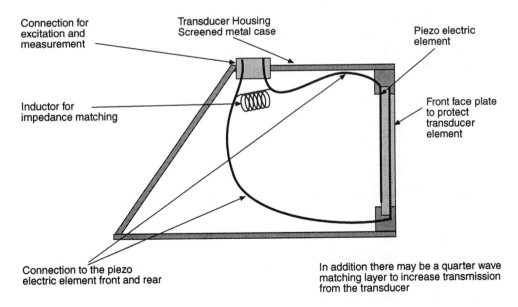

Figure 3.23 Transducer design

For imaging purposes the backing material is chosen to match the impedance of the transducer material. To avoid an excessive reflected pulse being returned by the rear face of the transducer and thus prolonging the emitted wave, the backing should provide high attenuation. The backing is loaded with tungsten powder.

4. Thickness mode vibrations at the desired frequency are obtained by choosing the element thickness to be half the wavelength at the resonant frequency.

5. For medical imaging purposes the object is required to be within the near field of the transducer since the field spreads after this point and therefore reduces the lateral resolution. Therefore having chosen the operating frequency and the required depth of field the disc diameter can be determined from equation (77).

6. The proportion of acoustic power transmitted from the active element to the tissue depends on the acoustic impedance match of the patient to the transducer element. Recalling the result for the transmission of sound through a thin layer in section 3.7.4.3, total transmission can be achieved if the impedance of a layer one quarter of a wavelength thick is chosen so that its acoustic impedance is equal to:

$$Z_2 = \sqrt{(Z_1 Z_3)}$$

where Z_1, Z_2 and Z_3 are the acoustic impedances of the transducer element, matching layer and patient respectively.

Therefore the acoustic impedance of a transducer element can be matched to that of the body and total transmission achieved. However, thickness of the layer will only equate to one quarter of the wavelength at a particular frequency. Therefore, although useful, matching layers tend to reinforce the narrow bandwidth characteristics of ceramic transducers.

7. The electrical impedance of the transducer element can be matched further to reduce the Q of the resonance and therefore increase the transducer bandwidth. This is achieved by using a shunt inductor to match the impedance of the transducer disc at its resonant frequency. The capacitance of the transducer disc can be determined from its thickness, the dielectric constant of the piezo electric material and the electrode area. The inductor required to match this at resonance is given by the following equation.

$$f = \frac{1}{2\pi\sqrt{LC}}$$

$$L = \frac{1}{(f2\pi)^2 C}$$

where f is the resonant frequency of the transducer and C is the capacitance of the transducer.

The impedance of the transducer may be matched to that of the receiving amplifier by a transformer.

8. The transducer housing protects the relatively fragile transducer element and provides electrical screening. The housing may also be shaped so that reflections from the back surface are not directed back to the piezo electric element.

4

Physiological Instrumentation

4.1. Introduction

The purpose of a Physiological Instrument is to provide information to a Physician as to the function and performance of an organ, group of organs or system within the body of a patient.

The performance may be a comparison of the function now to the function at some previous time or to the function of some other patient. The measurement may be specific to the individual or compared to a value defined for normal or abnormal populations.

Measurement	Signal Characteristics	Measurement Principle
Heart Rate	max = 300 bpm min = 25 bpm normal range = 69–90 bpm	Measurement derived from blood pressure measurement, pulse oximeter, ECG measurement or from direct heart sounds
Blood Pressure Arterial	max = 300 systolic min = 20 diastolic normal range = 120/80 bandwidth = 200 Hz	Measured with sphygmomanometer and stethoscope or strain gauge transducers alternatively invasively measured by catheter tube and pressure transducer
ECG Electrocardiogram	Voltage 10 μV to 5 mV normal peak reading 1 mV bandwidth = 1 kHz max	Measured by electrodes placed on the arms and legs or from chest electrodes
EMG Electromyogram	Voltage 20 μV to 500 μV bandwidth = 2 kHz max	Measured by electrodes placed close to the muscle group under examination or by needle electrodes implanted into that muscle
EEG Electroencephalogram	Voltage 2 μV to 200 μV bandwidth = 100 Hz max	Measured by electrodes placed on the scalp above the region of the brain under investigation

Figure 1 Common physiological measurements

The measurement may concern an individual organ such as the heart or the collective function of a group of organs all contributing to a particular effect.

Figure 4.1 shows a table of the typical size and type of signals

How can measurements be taken?

- Internal measurements (blood pressure by transducers introduced into the arteries).

- Surface measurements (electrical activity of the heart measured by potentials on the skin's surface).

- At a distance using signals which are emitted from the body (infra red, including sensing of transmission of radiation through the body).

What can be measured?

- Pressure, flow, velocity, force, acceleration, impedance, temperature, chemical concentration, gas concentrations.

4.2. Measurement Systems

The following sections are a brief introduction to the application of electronic measurement systems used in medical applications. This material is provided here only as a brief summary: the interested reader is referred to dedicated texts on these systems for a more thorough insight into current practice.

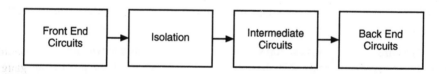

Figure 4.2 Generalised physiological measurement system

4.2.1. Transducer

A transducer is the functional element that converts one form of energy to another. A physiological measurement instrument may consist of a multitude of transducers and electronic circuits to process their signals. Transducers convert a parameter from one form of energy to another form which is amenable to measurement. For example pressure measurement transducers consist of a membrane which converts a pressure difference into movement or strain, which in turn is transformed into a varying resistance by a strain gauge. The resultant varying resistance can be incorporated in a circuit to enable measurement of relative pressure. There is a vast and ever increasing number of transducers available to the engineer: many employ semi conductor technology. Examples of transducers are: piezo electric crystal, strain gauge, and photo diode.

4.2.2. Front End Circuits

Signals originating from transducers often require conditioning before they can be analysed. Many transducers require excitation, as is the case with strain gauges, or incorporation within a circuit so that their characteristics may be analysed. The function of front end circuits then is to provide excitation of the primary transducer and to condition the detected signal. Signal conditioning may include pre-amplification and filtering.

4.2.3. Isolation

Modern physiological instrumentation is designed to rigorous safety standards to provide patient protection from the instrument or other inter connected instruments. Therefore there is a requirement to isolate the transducer and front end circuits from the rest of the equipment, to reduce the possibility of dangerous currents and voltages coming into contact with the patient. The isolation also serves to protect the patient from instrument faults. Isolation is provided as close to the patient as possible so that most circuits are separated: this reduces the demands on the power supply for the isolated section and the complexity of the isolation.

4.2.4. Intermediate Circuits

Intermediate circuits provide signal conditioning, filtering, amplification and analysis. For example the return signal from a Doppler blood flow detector is amplified in front end circuits, further amplified and filtered in the intermediate circuits. The signal is then modulated and the frequency deviation detected using either a zero crossing detector or phase locked loop. Thus the intermediate circuits provide signal conditioning, transformation and detection.

In modern medical equipment signal transformation and detection is often done digitally. Increasingly the detected signal is intensively processed to extract the information which represents the required measurement.

4.2.5. Backend Circuits

Backend circuits provide analysis, display and interpretation. Traditionally the most common form of display was a paper chart recorder, although this has largely been replaced by Cathode Ray Tube (CRT) displays. However, most equipment manufactured within the last decade has incorporated computer generated displays and there is now a move from CRTs to Liquid Crystal Display (LCD) technology.

4.2.6. Future Developments

The future may see computerised physiological measurement instrumentation developed that is part of an 'Expert System'. That is a system which mimics the decisions and experience of a skilled physician, which may well 'make decisions' to alter treatment and implement them. Chapter 7 provides a further description of their application in medicine.

4.2.7. Amplification

Elsewhere in this chapter, we quantify the signal levels seen when measuring biopotentials such as in either Electrocardiography (ECG) or Electroencephalography (EEG) measurement systems. The signals involved were between a few V and several mV. The impedance required of the measurement system had to be sufficiently high as not to degrade the signals of interest. The bandwidth of the signals in which we were interested was, however, limited.

The system gain required in order to drive either a CRT display or the motors used in a paper recorder must generate several volts from a relatively low impedance source. Gains of between 10^3 and 10^6 are therefore required. More exacting is the source impedance requirement which must be coupled with the need to provide safe isolation of the signals. The input impedance of the amplifiers may need to be between 10^6 and $10^9\Omega$. This level of impedance coupled with the need to measure very low bandwidth signals leads to the use of either MOSFET or chopper stabilised amplifiers.

The main problem associated with high gain directly coupled amplifiers is that they are inherently liable to amplify equally both the required signal and any instabilities that may be present in their first stage. The instabilities are primarily due to the effects of thermal drift of their first stage components. As an alternative to DC coupling, chopper stabilised amplifiers are used in which the signal to be amplified is sampled at the chopper's frequency, and the sampled frequency is itself multiplied by an AC coupled amplifier which does not feed forward the drift from each stage of the amplifier. An outline of the chopping process is shown in Figure 4.3 below.

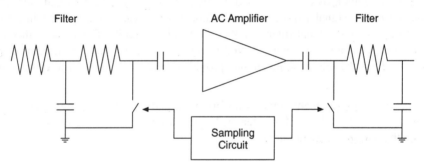

Figure 4.3   Chopper stabilised amplifier

4.2.8. Filtering

The signals obtained from biopotential measurements are frequently very specific in their form and of limited bandwidth. Whilst the DC component itself carries no information and is of little interest, we are often interested in low frequency signals. Additionally, the source impedance, as we noted in the last section is often high and the leads connecting small signals between the patient and measurement apparatus are necessarily long. They are therefore susceptible to both capacitive and magnetic pickup of electrical noise. A major source of electrical noise is often the power supply at a frequency of 50 or 60 Hz. There are also noise sources operating at radio frequency, such as electro surgical apparatus, communication equipment and radiated noise from switching equipment including power supplies.

Filters used in modern circuits almost always rely on the use of operational amplifiers as active components. The use of amplifiers obviates the need for the use of inductors, so that accurate filtering characteristics may be obtained with RC circuits which do not suffer from manufacturing difficulties. They also typically have low output impedances from each filtering stage which permits their use in tandem without coupling between separate filtering elements.

Unwanted signals may be removed from the required information to a large extent by selectively amplifying only the frequencies in which we are interested. Filters, however, vary in their types and complexity. Simple filters remove either high frequencies, low frequencies or pass only a range of frequencies, as illustrated in Figure 4.4: they are termed low pass, high pass and bandpass filters respectively.

The slope of a filter's characteristic attenuation as a function of frequency is controlled by the order of the filtering process. High order filters, whilst they provide sharp cut off, suffer from uneven transmission characteristics in their pass band.

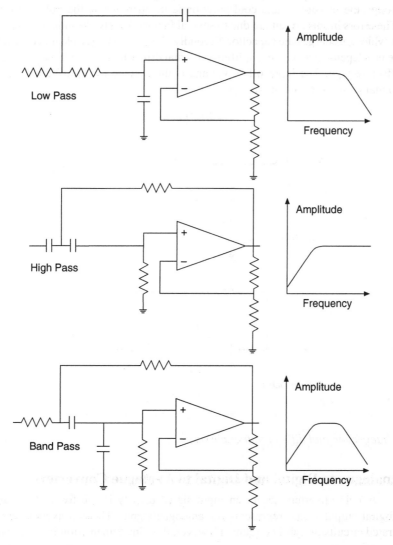

Figure 4.4 Active filters

Filters may either be implemented as analogue circuits or they can be built using digital components which analyse the frequency components of a digitised signal directly in the frequency domain and produce the desired cut off characteristic. The output of a digital filter may require to be reconstructed into its analogue form for later display.

4.2.9. Differentiation and Integration

Certain signals may require to be interpreted by either differentiation to examine their rate of change or be integrated, for instance to reduce certain noise characteristics. Both of these functions may be undertaken by analogue electronic circuits.

The integration process uses a circuit of the form shown in Figure 4.5. In this circuit the integrating element is effectively the capacitor. The capacitor is charged by the current which is fed through the resistor. The amplifier's inverting input is a virtual earth, so the current flowing through the resistor in this configuration is independent of the voltage across the capacitor. The errors in this circuit are due to several factors. Firstly, the amplifier has a finite bias current which discharges the capacitor. Secondly, the capacitor has a leakage current, and finally there is a dependence on the amplifier's thermal drift which is exhibited as a change in its input offset voltage. The input offset voltage is the voltage required between its input terminals to maintain a zero output voltage.

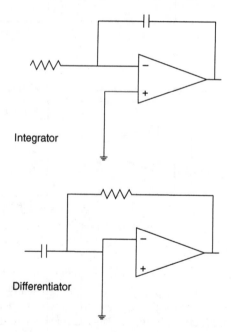

Integrator

Differentiator

Figure 4.5. Integration and differentiation circuits

4.2.10. Analogue to Digital and Digital to Analogue Converters

Analogue to Digital converters accept an input signal usually in the form of voltage and produce a digital output which represents the analogue signal. These converters are either single integrated circuits or hybrid circuits. However, their implementation is independent of their circuit configuration.

A conversion process must be undertaken if signals are either to be stored or processed by computers. The representation of signals in a digital form also permits their transmission through systems in which there would otherwise be serious concern about signal distortion or deterioration. This is particularly important in a medical environment as we frequently require to provide electrical isolation between measurement and processing circuits for patient safety. The normal forms of coupling circuit are either transformer or optically based. Transformer based circuits are normally difficult to build with adequate accuracy and are therefore expensive. Optical coupling methods do not readily lend themselves to the direct transmission of analogue signals as their transfer characteristics are normally far from linear. Both of these isolation methods may be used to transfer digital information where either non-linearity or the introduction of noise do not immediately compromise the signal accuracy.

The inverse conversion process clearly does not present us with these problems as it simply requires us to restore a digital representation of a signal into its analogue form.

4.2.10.1. Sampling Frequency or the Nyquist Rate

Several major factors characterise the digitisation process. Firstly, we note that if a signal is to be completely characterised by sampling, then the sampling must be carried out at a rate which is sufficient not to cause the loss of information. This rate, called the Nyquist Rate, is twice the maximum frequency contained in the signal. If sampling is carried out at a lower rate, then the sampled signal contains information derived from signals at above the frequency for which the sampling was valid, and permits them to be represented in effect as lower frequency signals. If the sampled signal's frequency spectrum is then analysed for its frequency components, it may reasonably contain only frequencies up to half the value of its Nyquist Frequency. The components of the signal at higher frequencies will still be represented there as energy, though their energy will falsely be indicated as lying within the permitted band. This error process is known as Aliasing. A mathematical presentation of the sampling process is given by Bracewell (1986).

4.2.10.2. Quantisation Error

A further error in the sampling process is in converting the sampled signal from its analogue samples into the numbers required for representation in a digital form. Any converter has a limited precision and a limited range of voltages which it may represent. When you specify an analogue to digital converter for use in an application, you will firstly need to note the voltage range of the signal in which you are interested and match this, possibly by using amplifiers, to the available converters. Secondly, you will be concerned to ensure that the conversion of the signal to a digital form does not lose required information. Clearly there is little purpose in attempting to discriminate between signal levels which are separated by less than the noise level in the signal you are measuring. However, conventional converters normally provide for conversions on a linear scale so the precision of measurement is greater for large signals than smaller ones. You will require a conversion process that caters adequately in terms of the precision of all signals in which you are interested. To achieve an adequate dynamic range for the conversion process may require you to use additional circuits which alter the gain of a preamplification stage and provide separate indication of the scale factor employed separately to a data acquisition process. In all cases, the 'quantisation error' introduced by the digitisation process is equivalent to the voltage represented by half of the least significant bit of the converter.

4.2.10.3. Types of Converter

Several types of circuit exist for both digital to analogue and analogue to digital conversion.

Digital to analogue converters are normally relatively simple devices in which each bit of the digital signal is used to control whether a reference voltage is applied to a resistor. The value of the resistor is chosen to correspond with the magnitude of the bit with which it is concerned: the current which flows in each resistor is therefore dependent on the applied digital signal. The currents may therefore be summed and amplified by an operational amplifier to generate a voltage which represents the number presented to the converter. The performance of the converter is controlled by the precision of the components used, and in particular their susceptibilities to thermal drift. The speed of the conversion process is controlled by the device's switching time, the settling time of the amplifier to an impulse and the capacitance of its input stage.

Analogue to digital converters come in a greater variety, depending on their application. A simple way to build a converter is to use a digital to analogue converter, a clock and control logic. This is shown in Figure 4.6. The major drawback in this case is that the conversion time is controlled by the clock rate, whose period must be greater than the converter's settling time times the precision of the conversion process. There are several variants on this arrangement which at the expense of additional circuit complexity accelerate the conversion time from that obtained by the simple process. The most straightforward of these uses the successive approximation technique in which attempts are made to estimate the magnitude of the analogue signal starting at the most significant bit of the converter. Each bit of the converter in turn is tested and the output is compared with the analogue signal. The bit is left set after the test if the generated signal remains below the level of the analogue signal. This conversion process can be carried out at each level at approximately the settling time of the converter, yielding a total conversion time of

Setting time × number of levels

To obtain very high conversion rates it is possible to convert all bits in parallel using high speed precision comparators together with a precision resistor network. The arrangement is shown in Figure 4.6. Whilst this method allows for the conversion of signals at many MHz, its complexity increases quickly with additional precision and places ever greater demands on component accuracy.

4.3. Transducers

4.3.1. Types of Transducers

The most useful form of transducer is one which converts energy to an electrical signal as this can be readily used and processed. Thus a thermocouple is more useful in instrumentation than a mercury in glass thermometer as the signal it produces is electrical and can be readily interfaced to other circuits for interpretation, display and recording.

Some transducers are used in a rather indirect manner. For example a displacement measurement transducer may be used to measure pressure. In this case a membrane is exposed to a pressure difference and its deformation is measured by a potentiometer.

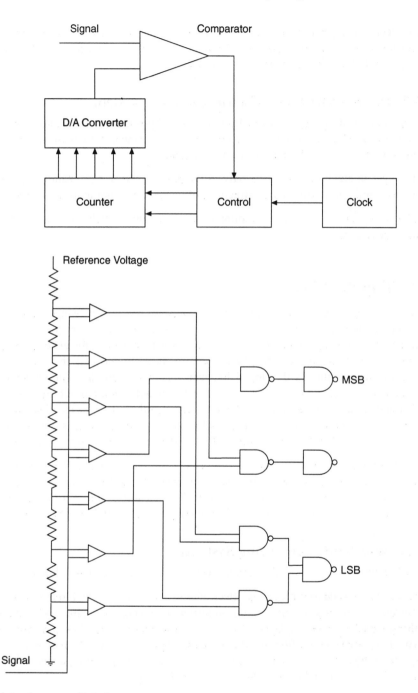

Figure 4.6 Analogue to digital converters

Some measurements are made by active intervention to obtain a signal. As an example injecting a small quantity of ice cold saline into the blood stream enables measurement of flow in the circulation system. Monitoring the temperature change at a point downstream determines the flow rate.

Transducers vary greatly in complexity. Modern transducers may be complicated semi-conductor devices which respond to chemical changes. Transducers should be designed to offer high reliability if their performance is safety critical.

4.3.2. Desired Attributes of a Measurement System

1. The amount of energy removed from a system by a measurement should be small. All systems are changed when a measurement is performed. A thermometer needs a finite amount of energy, in the form of heat, to function.

2. The system should be sensitive only to the desired signal. For example, ECG measurements should be sensitive to contractions of the heart and not to those of other muscles.

3. The measurement should be minimally invasive: the transducer should cause as little damage as possible.

4.4. Biopotentials

As we saw in Chapter 2, the depolarisation of muscles and nerves, which respectively cause contraction and the passage of information, is associated with the movement of chemicals across a semi-permeable membrane. This ionic movement causes the generation of an action potential. If two electrodes are placed near to an excitable cell then when the cell is depolarised a potential is developed between the two electrodes. Biopotential recording effectively measures the potential produced by cell depolarisation or the electrical activity of nerves and muscles. Essentially, cell membrane depolarisation is identical in nerves and muscles. The amplitude of the response is far greater in muscles: it is in the order of 1000 times greater from heart muscle contraction than through a neural nerve impulse.

The measurement of muscle contraction through the measurement of potential changes is called Electromyography (EMG), the measurement of the contraction of the heart, which can be viewed as a special type of muscle, is called Electrocardiography (ECG) and the measurement of nerve activity within the brain is termed Electroencephalography (EEG).

4.4.1. Biopotential Recording Systems

A block diagram of a typical biopotential recording system is shown in Figure 4.7.

To perform a biopotential recording, electrodes or leads are attached to the patient and then connected to a high gain differential amplifier with good common mode rejection. The resulting signal is then displayed. The system also incorporates low pass filtering of the signals and perhaps notch filtering at the mains supply frequency. As biopotential recording necessitates that a low impedance connection is made to the patient, patient safety has to be assured. This is achieved by isolating the pre-amplifier stages.

We are surrounded by power lines and equipment operating from a mains supply. This is particularly true in hospitals where much equipment is gathered together. Power lines operate typically at levels of hundreds of volts. The patient standing in a room in Figure 4.8 is capacitively coupled to the power lines. A biopotential measurement, such as an ECG, is then susceptible to a significant signal due to this coupling. As the biopotential is of the order of 1mV the capacitive interference may completely mask the biopotential signal. Early

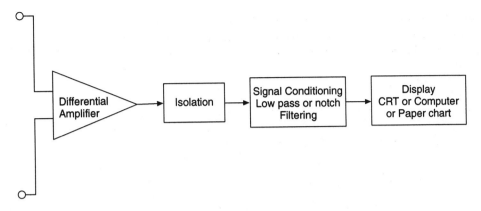

Figure 4.7 Generalised biopotential recording system

measurement systems were employed only in screened rooms in hospitals, and the recording equipment was battery operated. Now differential amplifiers which reject common mode signals facilitate this type of measurement in the presence of large interference signals.

Capacitive coupling of the power lines causes a signal at the first electrode which with respect to the common ground will be approximately the same as the signal at the second electrode. The differential amplifier amplifies the difference between the signals at the two electrodes. The ECG signal measured at one electrode is out of phase with the potential measured at the other electrode and therefore the difference signal is amplified. Hence the noise and power supply coupling are common to both electrodes and are therefore rejected. The ECG signal is different at the electrodes and therefore amplified. The ratio of the amplification of the required differential signal to the amplification of common mode signal is termed the common mode rejection ratio. For biopotential recording a high common mode rejection ratio is required. The common mode rejection ratio for biopotential recording apparatus is normally between 80 and 120 dB for ECG and EEG measurements. The impedance of the amplifier must be large so that the biopotential is developed across the input of the amplifier and not across the patient electrode interface. The input impedance is typically in excess of 5 MΩ.

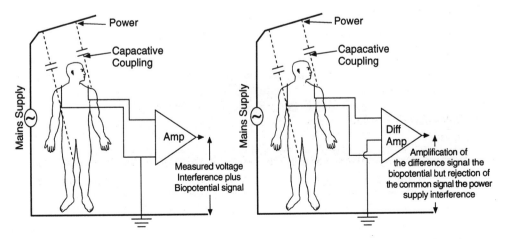

Figure 4.8 Amplifier configuration

The amplifier shown in the Figure 4.8 is referred to ground: this used to be a common configuration. As ground referenced equipment incorporates a conduction path to ground, the patient is at risk of electric shock in the event of a fault from the biopotential amplifier, other medical instrumentation or from ancillary equipment such as lighting and radios. Therefore modern biopotential amplifiers are constructed with isolated differential input sections and earth-free patient connections.

The leads from biopotential measurement equipment form a loop which potentially generates a current in the presence of a varying magnetic field. To reduce this effect the leads to a biopotential recording machine are twisted and screened.

A patient's skin resistance can vary between 100Ω and $2\,M\Omega$, depending on its condition. The resistance of thick areas of skin, like the soles of the feet, is higher than the resistance of upper parts of the body. Skin resistance is highly affected by moisture and, if the patient perspires, the resistance at the point of measurement will decrease drastically. The change in skin resistance with perspiration is made use of in the lie detector, where sudden changes in perspiration cause a drop in resistance, which is an indication of the patient's stress. Additionally, the patient's skin contains mineral salts. Therefore, to maintain a standardised low impedance interface with the patient, electrodes with a metal salt solution are used. This ensures that the skin is moist and the resistance stable and low. The metal salt sets up an ionically stable interface. If bare copper wires were used oxygen would be liberated at the electrode. With a silver chloride electrode, a balance redox reaction is maintained between chlorine and silver ions.

The epidermis is a semi-permeable membrane, so if a silver chloride electrode is placed upon the skin, chloride and silver ions can permeate the membrane. Thus a potential difference may be developed across the epidermis. Therefore, to prevent this happening, a patient's skin is prepared by rubbing it with an abrasive compound to strip the surface and stop a potential being set up across the semi-permeable membrane. A schematic of an electrode is shown in Figure 4.9. Any movement of the electrode on the skin potentially causes signals to be generated. Therefore, the electrode is designed to maintain an even flat contact under which no movement will take place.

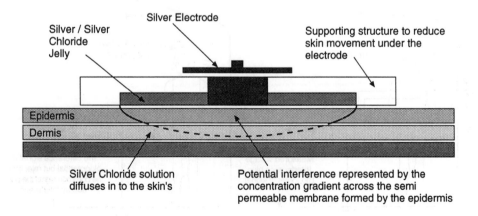

Figure 4.9 Biopotential electrodes

In addition to the electrodes described above, implantable electrodes are sometimes used. These employ a needle with electrodes mounted at its tip. The skin's resistance is broken and therefore the amplitude of the signal received is larger than that from surface electrodes. The use of transcutaneous needle electrodes also allows measurement near to the desired excitable cell. For instance, if a signal is required to be detected from a muscle fibre in the biceps the needle electrode can be placed in that muscle group. Surface electrodes detect an average signal originating from a large volume of the patient whilst needle electrodes record from a specific small volume around the needle tip.

4.4.2. ECG

The assessment of the functional behaviour of the heart by measurement of the potentials associated with cardiac muscle contraction is perhaps the most widely recognised biopotential recording.

The human heart can be considered as a large muscle whose beating is simply muscular contraction. Therefore contractions of the heart cause a potential to be developed. The measurement of the potential produced by cardiac muscle is called electrocardiology.

The depolarising field in the heart is a vector which alters its direction and magnitude through the cardiac cycle. The placement of the electrodes on the surface of a patient determines the view which will be obtained of that vector as a function of time. The most commonly used electrode placement scheme is shown in Figure 4.10. Here the differential potential is measured between the right and left arm, between the right arm and the left leg and between left arm and left leg. These three measurements are referred to as leads I, II, III respectively. This measurement lead placement was developed by Einthoven who stated that through measurement of lead I and lead II the signal seen at lead III could be calculated. This is the most basic form of ECG lead placement: from this the various features of the heart's depolarisation can be calculated. Clinically there is a range of lead placement schemes which incorporate limb leads and chest leads.

The heart consists of four chambers as we saw in Chapter 2, two atria and two ventricles. The atria contract together to force blood into the ventricles and then the ventricles contract.

A typical ECG trace is shown in Figure 4.11; the peaks and troughs of the waveform are labelled in accordance with normal medical practice. The first positive peak, the P wave,

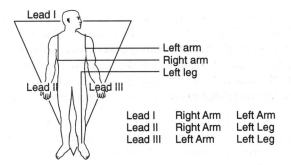

Figure 4.10 ECG lead conventions

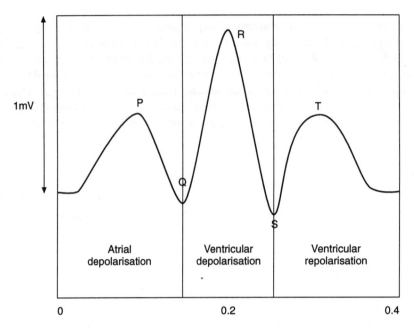

1mV

R

P

T

Q

S

| Atrial depolarisation | Ventricular depolarisation | Ventricular repolarisation |

0   0.2   0.4

Figure 4.11 A typical ECG trace

originates from the depolarisation of the atria. Following this there is a negative wave caused by the repolarisation of the atria. However, this waveform is usually masked by the depolarisation of the ventricles shown as the Q R S complex in Figure 4.11. The repolarisation of the ventricles causes the positive wave labelled T.

Therefore the ECG waveform shows the clinician the electrical waveforms associated with the contraction of the atria and ventricles. From an ECG a clinician may determine the relative timing of the contractions of the atria and the ventricles and assess the relative amplitude of the atrial and ventricular depolarisation and repolarisation. This information may allow the identification of mild heart block. Following a heart attack a patient's ECG shows changes as the timing and shape of the waveform are dependent on the transmission of the waveform through the muscle tissue. This changes with ischaemic muscle damage associated with heart attacks.

4.4.3. EMG

ECG measurement is the recording of the potential produced by cardiac muscle. Electromyography is the recording of the electrical activity of muscle: therefore ECG measurement is a type of EMG recording. EMG measurements are taken with essentially the same apparatus as is used for ECG measurement. However, in the majority of cases the signals received are smaller than those obtained in ECG recording as fewer muscle fibres are involved. The size of an EMG signal is directly related to the number of muscle fibres excited.

Clinically EMG is used to determine the function of muscle groups following trauma. It may also be used to assess muscle function following suspected neurological damage. EMG signals are larger than the signals associated with nerve depolarisation. EMG is therefore used to detect the arrival of motor nerve information, that is sensing the contraction of a muscle group following its excitation (see section 4.4.5.3).

4.4.4. EEG

Electroencephalography or EEG is the measurement of neural activity within the brain. Each neurone in the brain receives and transmits information through the depolarisation of its cell body sending an action potential along the nerve fibre. Within the human brain there is continuous activity and therefore continuous depolarisation and re-polarisation of neurones. This results in the continuous generation of electrical signals. These signals can be detected by electrodes placed on the scalp (see Figure 4.12). The measurement can be performed with instrumentation similar to that used for ECG measurement but with higher gain and common mode rejection. Signals are much smaller than ECG signals with an approximate amplitude of 2 µV. Usually surface electrodes are used of similar construction to ECG electrodes, with the metal-electrolyte interface being silver / silver chloride. During operations, needle electrodes may be implanted into various parts of the brain. This has allowed researchers to quantify which parts of the brain control which bodily functions.

We are continuously thinking and, therefore, EEG signals appear much like electrical noise. In certain circumstances, the brain waves or EEG signal may have characteristics which can be identified. If a person is relaxed with the eyes closed lying in a prone position, the brain waves form into a regular low frequency amplitude modulated wave form characterised as an 'Alpha wave' (see Figure 4.12). This occurs more readily in men than in women. The bandwidth of the signals is approximately 10 Hz. 'Beta waves' are classified as waveforms between 18 and 30 Hz. EEG is used in studying patterns of brain action during sleep to enable the quantification of a patient's sleep into classes such as rapid eye movement (REM approximately 30 second bursts of 8–12 Hz) and 'deep sleep' (frequencies of less than 4 Hz). During operations brain activity measured by EEG has been used to detect low oxygen and high carbon dioxide levels. Signals may also be recorded from electrodes implanted in the brain to assess the level of damage to a particular region.

A clinical use of EEG is in the diagnosis of epilepsy. Epilepsy is usually suffered by people during puberty or adolescence and may be triggered by fever or flashing lights. The EEG signals recorded during a fit are of higher frequency than normal. During an epileptic fit, EEG can be used to categorise the seizure as grand-mal or petit-mal as the characteristic EEG patterns in each case are different.

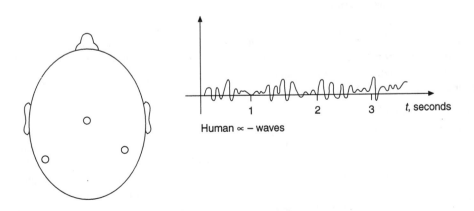

Figure 4.12 EEG measurements

4.4.5. Evoked Response Potentials

If you are sitting in a dark room, and suddenly there is a flash of light, then a signal is sent to the sensory cortex of your brain along the optic nerve. If your EEG is recorded then a small impulse is recorded, corresponding to the sensory cortex receiving the optical stimulation signal. In this instance the EEG electrodes are placed over the part of the skull above the visual cortex. This maximises the signal received from this area. However, much of the signal recorded is due to other stimuli and impossible to distinguish from the specific visual stimulus. However, if the light in the room flashed a number of times, and each time the EEG waveform was recorded, averaging the recorded waveform would diminish signals uncorrelated with the visual event. In this way, if a patient experiences a visual stimulation and the EEG is measured, background interference or neural activity can be removed to leave a signal corresponding to the receipt by the visual cortex of the signal along the optic nerve from the eye. The resulting signal is called a visually evoked response potential.

The EEG signal needs to be captured after triggering by the flash or visual signal. The visual signal may be a simple light flash, a change of pattern on a checkerboard display or a complex video picture. Usually more than 64 individual signals need to be averaged to obtain an adequate display. However, to discriminate the signal properly, about 200 or 300 visual signals need to be applied. To avoid cyclic events in the activity of the brain becoming correlated with the sensory signal, the signal is applied with a random time spacing between impulses.

Visual evoked response potentials have been used to try to determine the way in which the visual cortex processes information. However, the most common application of visual evoked response potentials is in the assessment of the reception of visual signals when patients are either unco-operative or unable to communicate. Visually evoked responses are therefore used to assess babies' sight. Visually evoked response potentials may help to distinguish between physical disability and malingering.

During brain surgery the optic nerve may be stimulated by flashing a light into the patient's eye by mounting LEDs to the patient's eye lid. The signals are recorded from inside the brain using needle electrodes. If the surgeon cuts close to the optic nerve, the nerve shocks and stops transmitting signals and therefore the EEG signal disappears. The surgeon then knows that the cut is close to the optic nerve.

4.4.5.1. Auditory Evoked Response Potentials

In the same manner that a patient's visual sensory system may be stimulated with a flash, a patient's aural system may be stimulated with a noise. The patient is either stimulated with a click or a tone burst (a sine wave of finite duration). The patient's brain activity is recorded using an EEG and, as with the visual sensory system, a number of EEG waveforms are recorded correlated with tone bursts or clicks.

The auditory signal may be applied to the patient using either earphones, headphones or loudspeakers. Due to the amplitude of the EEG signals (approximately 2 μV) loudspeakers are sometimes preferred. Owing to their greater separation from the point of measurement, they produce less electromagnetically induced coupling from the moving coil into the biopotential recording apparatus than do headphones. As with visual evoked response potentials, auditory evoked response potentials are used to assess industrial injury claims and suspected malingerers. They are also useful if the patient is too young or too psychologically impaired to communicate.

4.4.5.2. Sumato Sensory Evoked Response Potentials

The third main sensory system that we have after sight and sound is our peripheral nervous system. The nerves are concentrated in areas such as our hands and feet. They respond to pain, pressure and temperature. If they are stimulated a signal is passed from a peripheral nerve to our spinal column and hence to our brain. Therefore, if a patient's limb is excited and the EEG measured an impulse may be obtained in the same manner as for visual and auditory response potentials. However, the sumato sensory response potentials signals may be recorded at many places along the route. If, for example, a patient's hand is stimulated, the sensory nervous signal may be detected at points on the patient's forearm, shoulder and where the nerve enters the spine, as well as being detected in the EEG waveform. Sumato sensory evoked response potentials can be used to determine the extent of sensory sensation that a patient has in a limb or other part of the body. This may be particularly useful following a trauma such as a motor cycle accident, where the patient's nervous system may be shocked so that it blocks all signals. The patient may then be unable to feel anything from a hand or arm. Using sumato sensory response potentials the integrity in the nerve pathway can be assessed.

4.4.5.3. Measurement of Peripheral Nerve Velocity

The nerves which control the movement of a limb are called motor nerves. The nerve travelling down next to the ulna (one of the bones in the forearm) is called the ulna nerve. This nerve may be stimulated at a point close to the elbow as the nerve is close to the surface (see Figure 4.13). Using electrodes of essentially the same type as those used for ECG measurements, a current may be passed between the two electrodes through the patient. This excites

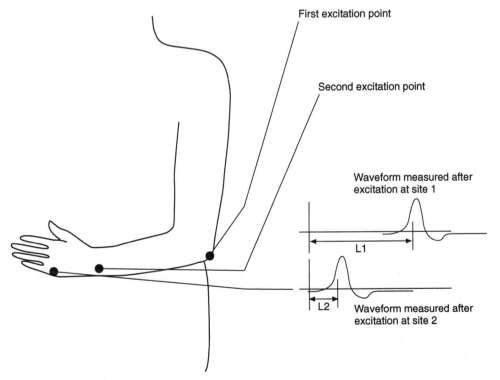

First excitation point

Second excitation point

Waveform measured after excitation at site 1

L1

L2

Waveform measured after excitation at site 2

Figure 4.13 EMG measurement

the nerve by depolarising its membrane and causing an action potential to be transmitted. The action potential travels along the nerve to a synapse, a junction between the nerve and muscle, in the thumb. An EMG then measures the electrical signal associated with the contraction of the muscle in the thumb. The time between nerve excitation and the muscle response as determined by the EMG is recorded as the latency. If the point of excitation is moved towards the wrist and the nerve excited again and the latency recorded as before, the difference between the two latencies corresponds to the time it takes for the nerve impulse to travel the distance between the two excitation points. In this way the peripheral nerve velocity can be measured. Measurement of peripheral nerve velocity may be useful in determining the extent of nervous damage following trauma.

4.4.5.4. Excitation of Nerves or Muscles

As we saw in Chapter 2, nerves and muscles are excitable cells. This means that their membranes' permeability changes if they are excited beyond a certain threshold. In a nerve, if a membrane is excited, then an action potential travels through the nerve and transmits information from one point to another. In a muscle, the travelling action potential causes a release of chemicals, which cause the contraction of the muscle. However, in both cases, the progression of information or the contraction of muscle is initiated by a change in membrane permeability. Change in membrane permeability can be measured using ECG, EMG or EEG methods as it is associated with a change in potential difference. Likewise, if a current passes through the semi-permeable membrane of a nerve or muscle, then an action potential can be generated. Thus the passage of current through the nervous tissue such as a muscle or a nerve, stimulates the tissue. Nerves and muscles may be electrically stimulated by placing electrodes on the skin and passing a current between them. The patient's skin is usually prepared in the same manner as for ECG electrodes and the electrodes are essentially the same as for that measurement. Electrical excitation of nerves and muscles can be extremely painful for the patient.

The duration of the nerve impulse is small. Potentials of up to 200 V may be used to generate the required current. Constant current sources are used rather than constant voltage as it is the passage of current through the nerve which causes its depolarisation. As with measurement of EMG or recording from peripheral nerves, needle electrodes may be used for stimulation. A small area, a particular nerve or muscle fibre group may be excited more accurately than with a surface electrode.

When recording or stimulating, surface electrodes pick up an average signal from a large tissue area, corresponding to a large group of nerves or muscles, whereas needle electrodes either excite or record from a small area or group of fibres.

4.4.5.5. Electrical Activity of the Heart

The heart is essentially a large muscular bag, which beats continually to pump blood around our bodies. The heart consists of four chambers. The two atria contract simultaneously, as do the two ventricles. The rate at which a heart beats is determined by the Sino Atrial node or SA node. A depolarisation pulse originating here spreads through both atria, causing contraction, and then spreads on to the AV node or Atrio Ventricular node. Depolarisation passing through the atria does not spread to the ventricles except by the pathway through the AV node. When the depolarisation pulse reaches the AV node it is delayed before passing to stimulate the ventricles. This allows ventricles maximally to fill with blood from the atria. The pulses then spread through the ventricles, causing their depolarisation and subsequent contraction. Thus the atria and the ventricles contract in a co-ordinated fashion. The rate at which the SA node stimulates the atria is determined by the para-sympathetic and sympathetic nervous systems as described in Chapter 2.

The SA node is the heart's natural pacemaker as it determines the rate at which the heart beats. Heart block is the condition where the electrical connection between the atria and the ventricles has been damaged, so that depolarisation does not spread to the ventricles. In this instance, the atria beat at a rate determined by the SA node and governed by the body's sympathetic and para-sympathetic systems. However, the ventricular contraction is uncoordinated with the contraction of atria. The ventricles then beat at a rate of around about 40 beats per minute. Even though this beating is uncoordinated with the atrial contraction, the ventricular contraction is sufficient to supply a life sustaining flow of blood throughout the patient's body. However, the patient will have very little energy and may faint and lose consciousness.

4.4.6. Pacemakers

A pacemaker is a device which takes over the timing control of the ventricular contraction from the body's natural system to ensure a rate fast enough to allow an active life for the patient. Pacemakers basically consist of a battery, a timing device and electrodes. The battery must be capable of supplying enough current to stimulate or excite the muscles in the ventricles and perhaps the atria for a number of years. As the heart beats approximately 70 times per minute, the battery life must be able to provide stimulation of $70 \times 60 \times 24 \times 365 \times$ the number of years it will survive. Therefore, the requirements for the battery are quite strict. Battery life is expected to be in the range of $3 - 10$ years. Usually 4 to 6 volt pulses are used to excite the heart with a duration of between 1.5 and 2.5 milliseconds. The electrical connection is made from the pacemaker to the patient's heart by leads. These must be rigorously designed as they must withstand a large number of contractions which result in flexing throughout their lifetime. The electrodes are either implanted in the heart or stitched to the surface of the heart. As a temporary measure, electrodes may be floated into the ventricles of the heart in the same way as a catheter is introduced during blood pressure measurement. The pacemaker itself is situated underneath muscle in the chest. The devices are about half the size of an audio cassette.

The third main block of a pacemaker is the timing device, which regulates when or if the patient's heart is excited. The first type of pacemaker was based around an asynchronous mode timer (see Figure 4.14). This constantly excites the patient's heart at a rate determined by the surgeon, but is approximately 70 beats per minute. The units are a simple design and therefore basic and cheap, but they suffered from an intrinsic problem. Since the ventricles sometimes spontaneously contract, if the pacemaker excites ventricles after their contraction, the patient's heart could fibrillate.

Ventricular Fibrillation is a mode of contraction in which the ventricles continue to be excited but in an uncoordinated fashion, causing the heart to flutter rather than to pump. In this state the patient receives no blood flow and therefore is in danger of dying. Once the patient's heart is in ventricular fibrillation, a large signal is required to depolarise the whole of the ventricles simultaneously to synchronise their depolarisation. Asynchronous mode pacemakers associated with heart attacks due to the competition between the pacemaker and the na spontaneous contraction of the ventricles themselves.

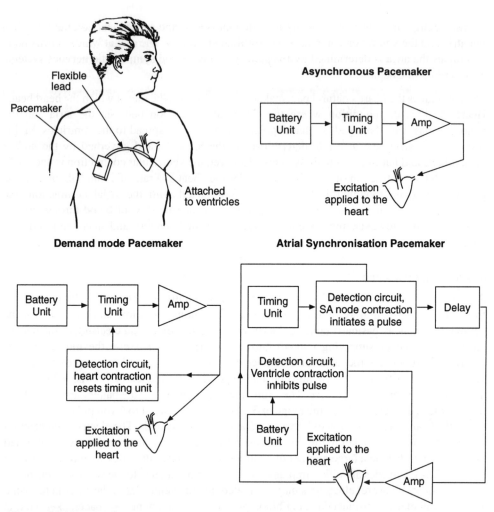

Figure 4.14 Pacemakers

The battery life of a pacemaker may be monitored non-invasively. Asynchronous pacemakers were designed so that their rate would drop as the battery level fell. For example, when the batteries were fully charged the rate of the pacemaker may have been 70.9 beats per minute, whereas two to three years later, when the battery power dropped, it may have dropped to 70.1. In this way the patient's general practitioner could monitor the battery life of the pacemaker non-invasively during a normal clinic.

A more sophisticated type of pacemaker is the Demand Mode pacemaker (see Figure 4.14). This type incorporates detection circuitry. Either separate leads, or those used to excite the ventricles, determine whether the ventricles have contracted. The pacemaker excites the patient's heart if the heart itself has not spontaneously contracted after a pre-determined period. Therefore, the demand mode pacemaker ensures a minimum heart rate but will not excite the patient's ventricles following a spontaneous contraction.

The third and most sophisticated type of pacemaker is the Atrial Synchronisation pacemaker (see Figure 4.14). In this type of pacemaker, separate electrodes are used to detect the contraction of the atria. The pacemaker introduces a small delay which models the delay in a healthy heart by the atrio ventricular node and then excites the ventricles. As with the demand mode pacemaker, the ventricles are not excited if spontaneous contraction occurs. However, in normal operation, contraction of the ventricles is synchronised with the contraction of the atria to ensure an efficient beating process. Since the excitation of the ventricles is determined by atrial contraction, itself controlled by the SA node, the body's own natural pacemaker, an atrial synchronisation pacemaker ensures that the patient's heart rate responds to the demands of the body. Thus if the patient is exercising, the body releases chemicals which cause the heart rate to rise. The rising rate of atrial contraction causes a corresponding rising rate of ventricular contraction, so a patient may lead a normal active life.

Both the demand mode pacemaker and the atrial synchronised pacemaker offer improved performance and reduce the likelihood of patients suffering from ventricular fibrillation due to mistimed excitation pulses. However, both these types of pacemaker can suffer in the presence of noise. The reader will be aware that in the entrance to many libraries and some shops there are signs saying that people wearing pacemakers should contact the staff before entering the building. This is to avoid interference from electrical equipment disturbing the natural function of the pacemaker. Microwave ovens are another source of electrical interference which may disrupt pacemakers.

The most advanced pacemakers sense the level of background interference and, if this is high, they fall back to an asynchronous mode operation, returning to their atrial synchronisation or demand mode when the noise level has dropped.

4.5. Blood Pressure Measurement

4.5.1. Medical Aspects of Blood Pressure Measurement

The measurement of blood pressure is one of the checks that your doctor routinely performs. Problems with the cardio-vascular system are often mirrored by abnormal blood pressure readings. More specifically malfunction of the heart can be directly diagnosed from blood pressure measurement. In the hospital environment blood pressure is monitored during operations and continuously recorded in intensive care units. Measurements of blood pressure within the heart may provide information about the integrity of the heart valves.

The normal resting patient will have a heart rate of approximately 70 beats per minute. Each stroke ejects about 70 cm^3 of blood from the left ventricle into the aorta (main artery leaving the left ventricle). At the hiatus of each stroke the pressure within the arteries reaches a maximum and then declines during the rest of the cardiac cycle. The peak blood pressure is referred to as systolic whilst the minimum value is the diastolic. The normal patient will have a systolic blood pressure of approximately 120 mm Hg and a diastolic blood pressure of approximately 80 mm Hg.

The body's circulation can be divided into two circuits: the Systemic circuit and the Pulmonary circuit. The pressure in the systemic circuit is high as the blood leaves the heart however, having passed through the body the pressure in major veins is as little as 4 mm Hg. The pulmonary circuit operates at relatively low pressure as the function of the lungs is to exchange oxygen. Pulmonary circuit pressure has a systolic peak of 25 mm Hg and a diastolic value of 10 mm Hg. Therefore blood pressure is pulsatile and dependent on the point of measurement. Clinically the doctor is interested in the systolic and diastolic reading and also the average pressure.

The heart rate is not directly related to the blood pressure as the impedance offered by the body is not constant. For example a sudden fright like finding that your alarm clock has broken on the morning of an exam will have a physiological effect known as the 'fight or flight' response. Following a sudden excitement the blood vessels near the skin's surface contract, therefore, reducing the flow of blood and minimising the blood loss from a superficial wound. Blood is also diverted from circulating round the stomach and intestines and instead lies in enlarged arteries and veins. The composition of the blood also changes, increasing the clotting speed, with corresponding viscosity changes. Therefore, the blood pressure increases regardless of any change in the heart rate. Using the analogy of relating the blood pressure to voltage and the flow rate to current and the circulation system to impedance. The blood pressure is clearly dependent on the heart rate, the flow rate and the impedance characteristics of the circulation system.

4.5.2. Sphygmomanometer Blood Pressure Measurement

The Sphygmomanometer is the most common method of blood pressure measurement. As it is non invasive it can be performed by your General Practitioner.

The measurement is performed by wrapping an inflatable cuff around a patient's upper arm. The cuff consists of a bag contained within a cloth sleeve and can be secured by a tie or velcro strap. The cuff is connected to a hand bulb pump and a mercury manometer. When the cuff is inflated, the pressure increases and the tissues in the patient's upper arm become compressed occluding blood flow in the brachial artery (see Figure 2.24a). The physician listens to the flow of blood at a point below the bag with a stethoscope. The pressure in the cuff is increased above the point where the blood stops flowing in the artery.

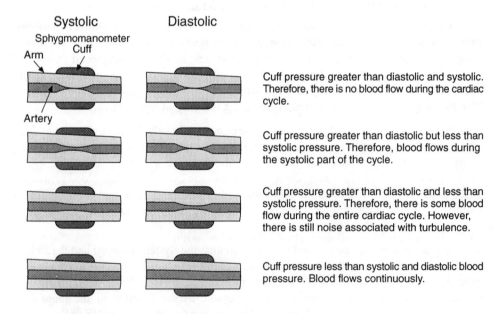

Figure 4.15 Sphygmomanometer cuff

The physician then gradually releases the pressure in the cuff; at a point when the pressure is equal to the systolic blood pressure blood begins to flow again. At this pressure blood squirts through the compressed artery during the systolic part of the cardiac cycle, but at the diastolic part of the cycle the artery closes again and no blood flows. Therefore, the artery opens and closes causing a characteristic knocking noise known as Korotkoff's sounds. At the point when these noises begin the clinician notes the manometer pressure as being equal to the systolic pressure.

The pressure within the cuff is then further reduced and consequently the artery remains open for an increasing portion of the cardiac cycle. When the cuff pressure is equal to the diastolic pressure the artery remains open during the whole cycle and the noises stop. The physician will record this cuff pressure as the diastolic pressure.

This method of measuring blood pressure suffers from several drawbacks.

- The measurement is subjective, that is to say it is dependent on the opinion of the clinician when the Korotkoff's sounds begin and end.

- The fidelity of the stethoscope used and of the hearing of the clinician directly control the measurement accuracy.

The sphygmomanometer is therefore a relatively inaccurate subjective measurement of blood pressure. However, it is non invasive, easy to perform and the instrumentation required is cheap. This is the type of blood pressure measurement performed in the front line of medical practice, the fault finding level. Measurements with greater accuracy require invasive techniques.

There is a number of different methods of determining the point at which blood begins to flow again other than detecting the Korotkoff sounds. As before, pressure in a cuff cuts off the blood flow. As pressure is decreased the blood flow resumes when the cuff pressure is approximately equal to the systolic pressure. This flow may be detected by a Doppler ultrasound blood flow detector (see Section 3.9.3). The Doppler device produces an audible output which is proportional to the blood flow velocity.

A transducer may be used to measure vibrations in the cuff which recommence as the cuff pressure approximately equals the systolic blood pressure. The vibration in the cuff reaches a maximum at a point when the pressure in the cuff is equal to the average blood pressure, and decreases and disappears when the cuff pressure is equal to the diastolic pressure.

Instruments which perform non invasive measurement of blood pressure have been developed which use both the vibration principle and the detection of Korotkoff's sounds. In both cases the cuff is automatically inflated and the vibration signal detected by transducer located in the cuff. These machines provide measurements of the systolic, average and diastolic blood pressures. However, these machines perform badly in the presence of other vibration signals.

4.5.3. Invasive Measurement of Blood Pressure.

Invasive blood pressure measurement allows more accurate blood pressure measurement, dynamic measurement and measurement of pressure at specific points in the circulation system.

4.5.3.1. Catheter Measurement

Invasive measurement of blood pressure is routinely performed using a catheter transducer. An incision is made in the patient and a catheter is introduced into the circulation system. A catheter is an open ended tube used to couple the pressure inside the patient to an external transducer: in other circumstances catheters may be used for drug infusion or drainage.

The catheter may be introduced into either the arterial or the venous system depending on the point at which the measurement is to be performed. The most common points of incision are in the neck, arm or groin, as these sites have large veins and arteries which are close to the skin's surface.

The catheter is filled with saline solution and connected to an external pressure transducer (see Figure 4.16). The saline filled catheter transmits the pressure at the catheter tip to the external transducer. The housing of the pressure transducer allows the catheter to be regularly flushed with saline solution to prevent blood from clotting at the open end of the tube (clots in the circulation system can obstruct blood flow and cause brain damage or damage to lung function). The transducer housing also allows the connection of a reference pressure source to allow for calibration of the transducer.

Once introduced into a blood vessel the catheter can be fed along the vein or artery reaching deep locations in the circulation system such as the heart or major arteries. Therefore the catheter tube has to be very compliant to bend around the structures within the body. The requirement for compliance contradicts the tube's primary function of transmitting pressure to the external transducer. If the tube is flexible and compliant then the pressure changes which are being measured may extend the tube and consequently be distorted when they arrive at the transducer. A further problem with catheters is that foreign matter or bubbles

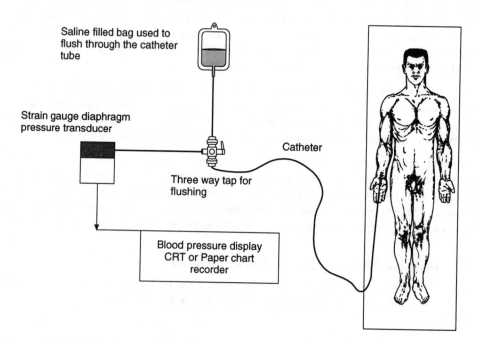

Figure 4.16 Catheter blood pressure transducer

within the tube may also degrade performance. Bubbles within the catheter compress when pressure increases and distort the pressure measured at the transducer. These factors limit the dynamic performance (i.e. bandwidth) of the catheter tube blood pressure measuring system. The bandwidth of such a system may be as little as 12 Hz.

However, the catheter system is relatively cheap, and the transducer is reusable and robust.

4.5.3.2. Catheter Tip Transducer

Another type of invasive blood pressure measuring system is the catheter tip blood pressure transducer. In this instance the transducer is mounted at the tip of a tube inserted into the patient and therefore directly measures the blood pressure at that point. The dynamic performance of this system depends on the transducer's bandwidth which may extend to as much as 1 kHz. Hence this kind of measurement is preferred when measuring the wide bandwidth signals found in the aorta and within the heart, particularly if associated with valve abnormalities. The intravenous tip transducer may be a little less compliant than the catheter system. However, its major drawbacks are price and operating life. The design requirements to provide an ultra miniature transducer make this kind of instrument expensive and it is easily damaged both in normal use and accidentally. The transducer's performance is also degraded by repeated sterilisation at high temperature.

4.5.4. Design of Blood Pressure Transducers

The type of transducer most commonly used in both catheter tube and intravenous blood pressure measurement systems is the strain gauge transducer.

4.5.4.1. Strain Gauge Pressure Transducers

If a wire is stretched the resistance measured between its ends increases due to a number of factors. The diameter of the wire decreases, the length of the wire increases and the resistivity increases. These three factors are related to the resistance of a wire by the equation given below.

$$R = \frac{\rho l}{A}$$

where A is the cross sectional area, l is the length of the conductor and ρ is the resistivity of the conductor.

In most strain gauges the changes in dimension are the dominant reason for the measured change in the resistance of a wire under tension. For small strains (those with changes of dimension of about 2%) the change of resistance with strain is approximately linear. If a wire is connected between two points monitoring the resistance of the wire correlates with the strain it experiences. This type of strain gauge is referred to as an un-bonded strain gauge and is uncommon in transducers today.

A bonded strain gauge consists of a plastic substrate upon which a layer of conducting material has been laid. The conducting material is laid out to maximise the length of the conductor in the direction of the required measurement. The bonded strain gauge may be glued to a material such as a block of steel. Any strain experienced by the block will be passed on to the bonded element. Bonded strain gauges can be fabricated from a variety of materials.

The temperature dependence of resistivity is possibly a limitation in the application of certain strain gauge transducers.

In a pressure transducer a membrane separates the fluid under test from a fluid at a reference pressure (normally air at atmospheric pressure); if the pressure of the test fluid exceeds that of the reference fluid then the membrane distorts in the manner shown in the Figure 4.17a. There is therefore a strain in the membrane proportional to the membrane's displacement and thus the pressure difference.

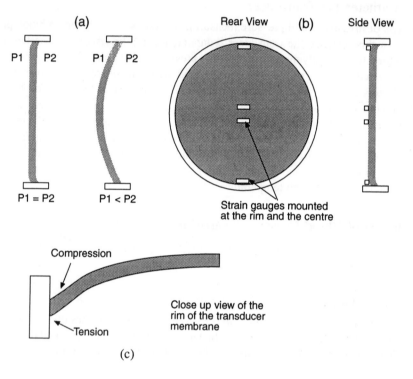

Figure 4.17 Transducer diaphragm

Strain gauges are mounted directly to the reverse side of the membrane. The resistance change measured will be extremely small and so gauges are incorporated in a Wheatstone bridge circuit. To maximise the sensitivity of the transducer and also to reduce the effect of temperature changes four strain gauges are used in a full bridge configuration. The transducer mounted in the centre of the diaphragm (see Figure 4.17b) experiences tension during membrane distortion and the gauges at the rim experience compression. This may at first seem rather surprising that the rim gauges will experience compression but if the situation is examined closely as in Figure 4.17c, then the membrane at this point can be considered as a beam under load and the situation understood.

Both catheter tip transducers and the comparatively large transducers used for catheter blood pressure measurement are constructed from a flexible diaphragm mounted with a strain bridge.

There are other types of pressure transducer which use other physical properties such as the piezo electric effect, capacitive sensing of membrane displacement and optical fibre techniques.

4.5.4.2. Safety of Blood Pressure Transducers

Blood pressure transducers are electrically hazardous as they penetrate the skin and therefore form a low impedance connection to the patient. The most susceptible organ to electrical accidents is the heart; this is measured directly by blood pressure transducers. The fluid filled catheter method establishes a conducting path to the heart and the intravenous transducer actually incorporates an electrical circuit within the patient's heart. On the other hand, the optical method may be completely isolated. For this reason optical transducer methods for blood pressure measurement may become popular.

4.5.5. Measurement of Blood Pressure within the Heart (the Swan Ganz Catheter)

In certain clinical conditions such as heart valve failure it is necessary to measure the blood pressure within the heart. This is achieved by introducing a catheter transducer into a vein or artery and feeding the catheter up against the flow. The patient is examined radiologically to determine the position of the catheter. If the catheter is introduced into an artery then it enters the left hand side of the heart. If the pressure transducer is introduced into a vein then it enters the right hand side of the heart.

4.5.5.1. Heart Catheterisation

To catheterise the right hand side of the heart a catheter is introduced into a vein and fed against the direction of blood flow. The position of the catheter is monitored by X rays as it is radio opaque. The transducer emerges at the entrance to the right atrium, and a pressure measurement may be taken at this point (see Figure 4.18). The transducer can then be pushed through and into the atrium where again a pressure measurement may be performed. Careful manipulation may then position the catheter in the ventricle and the pressure can likewise be measured at this point. The catheter cannot be manipulated into the pulmonary artery and so to take measurements there a balloon is inflated close to the end of the catheter which is

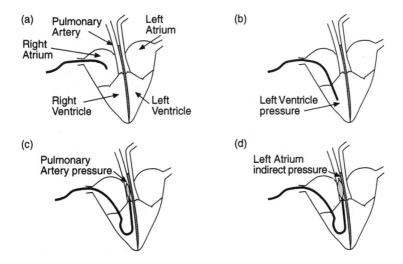

Figure 4.18 Heart catheterisation

carried by the blood flow through the pulmonary valve and into the artery. Once in the artery the balloon is deflated and the pulmonary artery pressure taken. With the catheter in this position it is also possible indirectly to measure the pressure within the left atrium. This is achieved by reinflating the balloon at the end of the catheter until blood flow in the artery stops. If there is no flow then the pressure within the pulmonary artery equals the pressure in the left ventricle as there is effectively a closed path.

This special process is performed using a Swan Ganz catheter. In special care wards patients are intensively monitored. Therefore patients are catheterised and the blood pressure within their hearts is continuously monitored. In addition the flow rate through the heart can be assessed by thermal dilution. This method comprises injecting a small quantity of ice cold saline solution into the heart and monitoring the length of time for this temperature fluctuation to be passed to a further part of the circulation system.

4.5.6. Measurement of Oxygen in the Blood

The function of the cardio-vascular system is to pass blood through the lungs so that haemoglobin can combine with oxygen, which can then be carried around the body. The measurement of the amount of oxygen in arterial blood is a direct measure of the performance of the cardio-vascular system in general. There are two clinically used methods of measuring the amount of oxygen in the blood: one relies on an optical measurement procedure, the other on a chemical reaction.

4.5.6.1. The Oximeter

Beer's Law relates the absorption of light passing through a solution to the concentration of the solution, the optical path length and the absorption coefficient. If the concentration of a solution is doubled, the absorption doubles and likewise an increase in the path length causes a corresponding increase in the absorption. This relationship is exploited in oximeters to measure the amount of oxygen in blood.

The active cells in blood which carry oxygen to tissue are haemoglobin. The spectral characteristic, the strength of absorption at a particular frequency, of haemoglobin changes when it is combined with oxygen. The spectral characteristics of oxygenated and deoxygenated haemoglobin are shown in Figure 4.19. The characteristic absorption functions cross at the isobestic point. At this point, the absorption of haemoglobin and oxyhaemoglobin are equal - this wavelength is 803 nm. The spectral absorption characteristics of blood account for the difference in the appearance of arterial and venous blood. Arterial blood (predominantly oxyhaemoglobin) is red whilst venous return (haemoglobin) is blue.

The oximeter determines the percentage oxygenation of haemoglobin as defined below.

$$PO_2 = \frac{Oxyhaemoglobin}{Oxyhaemoglobin + haemoglobin}$$

A sample of arterial blood taken from the patient is placed in a cuvette. A cuvette is a small rectangular glass vessel with known optical properties. The measurement is performed by measuring the absorption at two wavelengths, namely at the isobestic point and at 605 nm. As the path length (the dimensions of the cuvette) and the absorption coefficient are known the

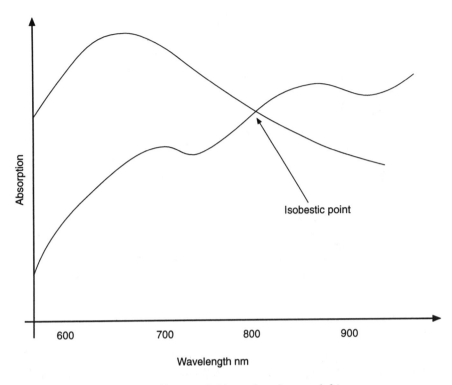

Figure 4.19 Optical properties of haemoglobin and oxyhaemoglobin

measurement at 803 nm allows the calculation of the total concentration of haemoglobin in the sample. When this is combined with a measurement at 605 nm equations can be solved to yield the percentage of oxygenated haemoglobin.

The advantage of this method of measuring the percentage saturation or the percentage oxygenated haemoglobin is that it is relatively easy to perform. The disadvantage is that a sample of blood has to be removed from the patient to perform the measurement. This means that it is relatively slow.

4.5.6.2. Oxygen Association

Oxygen in the blood is held there in two states:

1. Physically dissolved in the fluid.

2. Combined with haemoglobin.

Figure 4.20 shows the oxygen dissociation graph for blood. It shows on the abscissa the percentage of oxygen combined with haemoglobin whilst the ordinate shows the total oxygen in the blood. Initially the graph is relatively linear however, once the percentage of combined haemoglobin exceeds 80%, the graph levels off and a large increase in total oxygen correlates with a small change in percentage oxygenation. For instance, the 20% increase in the total oxygen from 8 Pa to 10 Pa on the graph is equivalent to a 2% change in the percentage oxygenation. Therefore, when using an oximeter, the percentage haemoglobin is a good judge of the amount of haemoglobin which is combined with oxygen. However, it does not directly

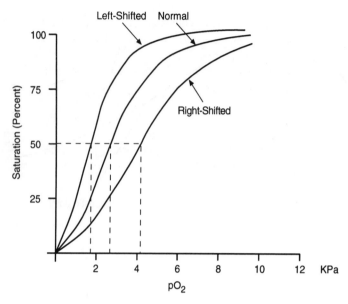

Figure 4.20 Oxygen disassociation curve

reflect the amount of oxygen in the blood; a small change, say from 98 to 96% in the haemoglobin concentration, can mask a change of 20% of the total oxygen within the blood.

A further problem is that some diseases or conditions cause a change in blood pH. This affects the oxygen dissociation curve, shifting it to the left or right, as shown in Figure 4.20. In these instances the relationship between the percentage oxygenation of the blood or the percentage oxyhaemoglobin to total oxygen in the blood is altered. Therefore, an oximeter measurement of blood oxygenation gives a good indication level of combined oxygen in the blood but not an accurate reading of the total oxygen present. Normally, the anaesthetist will expect a reading above 90% for the percentage saturation.

A further point of interest is that blood is composed of haemoglobin and abnormal haemoglobin, which has different spectral properties to normal haemoglobin. This may render the calculation of percentage oxygenation inaccurate. An oximeter measurement may also be disturbed if the patient is suffering from carbon monoxide poisoning, as this will alter the spectral characteristics of the blood.

4.5.6.3. The Pulse Oximeter

Oximeters rely on taking samples of blood, and placing them in external containers so that optical characteristics can be determined. This method is therefore slow and quite difficult to perform during an operation. The pulse oximeter uses essentially the same theory as the oximeter, but is a non-invasive measurement which can give a continuous reading of the percentage oxygenation of the haemoglobin.

A pulse oximeter takes a reading of the absorption for two wavelengths of light, as before, but through a portion of the patient's body containing arterial blood. The portions of the patient's body normally chosen are the ear lobe or the finger. A spring clip containing an optical source and detector is fastened around the patient's finger or ear lobe. The absorption through the patient is then dependent upon both the oxygenated and deoxygenated haemoglobin within the ear or finger, the tissues within the ear or finger and the skin, and in addition, the absorption of venous blood in the pathway. The light source emits pulses of light at both frequencies of light used in the measurement. A reading is also taken when the light source is off to determine the background noise or 'dark reading'. The pulses are emitted continually and so the absorption is continuously monitored. Due to the pulsing of the arterial blood, absorption is not constant with time as shown in Figure 4.21. The relationship between the pulseatile component of the absorption and the percentage saturation of haemoglobin has been empirically determined. The pulse oximeter, therefore, performs a continuous in-vivo measurement of blood oxygen saturation. However, it suffers from the same limitations as the oximeter.

Measurements performed by oximeters can also be widely affected by electrical interference. If a surgeon uses a diathermy in the operating theatre, then the pulse oximeter reading may be disrupted.

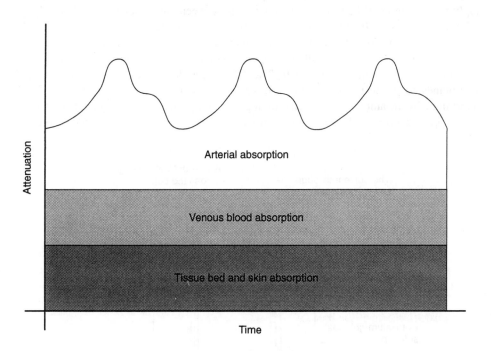

Figure 4.21 Tissue absorption

4.5.6.4. New Developments in Pulse Oximetry

To overcome the interference which pulse oximeters experience from other electrical equipment in the operating theatre, pulse oximeters are being built with the optical source and detector housed in a screened instrument case and the light fed to the patient by an optical fibre. A further modification to increase the accuracy of the calculation for measuring oxygenation is to include more than two wavelengths of light for measurement.

4.5.6.5. Chemical Measurement of Blood, Gas Content

The chemical oxygen sensor shown in Figure 4.22 can be used to measure the level of oxygen in a sample of blood or gas. The cell consists of two electrodes in a solution and a semi-permeable membrane. One of the electrodes is a silver / silver chloride electrode, whilst the other is platinum. The electrolyte within the cell is potassium chloride. The bottom of the cell is a semi-permeable membrane. There are two chemical reactions which can take place within the cell: one is silver combining with chlorine to form silver chloride and the other is oxygen combining with water to form hydroxyl ions.

$$O_2 + 2H_2O + 4e^- \rightarrow 4OH^-$$

$$Ag + Cl^- \rightarrow AgCl + e^- \tag{1}$$

The platinum electrode acts as a catalyst. The second reaction is oxygen dependent. For the water to dissociate to hydroxyl ions, oxygen needs to be present. The cell is operated by applying a potential of between 0.2 and 0.6 volts between the two electrodes. Initially, as the voltage is increased, the current increases between the two electrodes. However, it reaches a point when there is no increase in current with increased voltage. At this point the current that flows from one electrode to the other is dependent on the rate of the reaction and therefore the amount of oxygen entering the cell. The Clark Cell has specially designed permeable membranes to allow measurement of oxygen in blood. The reaction rate in the cell is determined by the amount of oxygen crossing the semi-permeable membrane from the blood. The resulting current in the cell is proportional to the amount of oxygen in the blood.

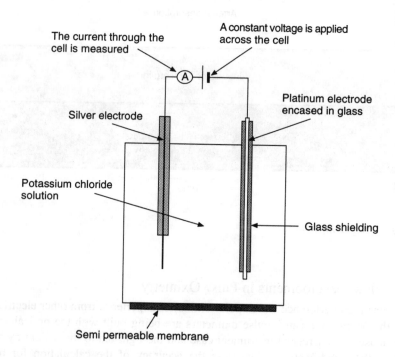

Figure 4.22 Chemical oxygen measurement

This cell measures the total amount of oxygen in blood. The dissolved oxygen and that combined with both normal and abnormal haemoglobin are measured. However, the cell cannot be used non-invasively. Either a sample of blood must be removed from the patient and placed on the semi-permeable membrane, or the cell itself must be miniaturised so that it may be inserted into the patient.

However, a cell has been designed using the same procedure as the Clark electrode for non-invasive blood oxygen measurement. This cell is placed on the patient's skin and the skin is heated. Heating the skin increases peripheral arterial circulation and opens the capillary pathways. Therefore, using a semi-permeable membrane especially designed for skin contact, the rate of the reaction is controlled by oxygen diffusing through the patient's skin from the arterial circulation into the cell. In this way the cell can be used to measure in-vivo oxygen concentration of blood.

4.5.6.6. Gaseous Oxygen Measurement

A further measurement of the cardio-vascular system can be obtained by measuring the oxygen inhaled and exhaled by the patient. Oxygen is a paramagnetic substance. Therefore, the presence of oxygen increases a magnetic field. Diamagnetic materials experience a force in a non-uniform field. The gaseous-paramagnetic oxygen analyser uses these two phenomena to measure the amount of oxygen in a sample of air.

The instrument is fabricated with two dumb-bells containing nitrogen, a diamagnetic substance, within a magnetic field (see Figure 4.23). The dumb-bells are suspended from a wire so that the magnetic field causes a torque, twisting the wires attached to the dumb-bells. At a point there is a stable position reached when the torque acting on the dumb-bells due to the magnetic field equals the torque exerted through twisting the wire. If the magnetic field were increased then the dumb-bells would rotate further.

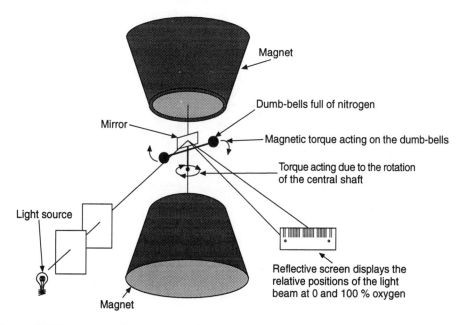

Figure 4.23 Paramagnetic oxygen analyser

The instrument is calibrated by filling the chamber full of inert gas and marking the displacement of the dumb-bells. If the chamber is then completely filled with oxygen, the magnetic field increases and the dumb-bells rotate further. This new position is noted. There is a linear relationship between the dumb-bell rotation between the first point measured with inert gas and the second point, measured with 100% oxygen filling the cavity. The amount of oxygen in the gas can be determined. The measurement is practically performed in a similar way to a spot galvanometer, in that a minute mirror is placed in between the two dumb-bells at the point of intersection with the wire. A light is shone onto the mirror and reflected onto a screen. The position of the reflection corresponding to zero oxygen concentration in the chamber and 100% oxygen in the chamber are marked as being 0 and 100% oxygen. Then the position of the reflection between these two marks can be related to the percentage in the sample.

Modern gaseous oxygen analysers incorporate a coil of wire around both dumb-bells so that when a current passes through the wire, an equal and opposite force to the magnetic torque due to the diamagnetic nitrogen is produced. In this way the current can be varied through the coils to maintain zero displacement of the dumb-bells. Hence, a graph of current through the dumb-bells to maintain zero position against oxygen concentration can be used to calibrate the device. The gaseous oxygen analyser malfunctions in the presence of significant concentrations of other paramagnetic gases, such as carbon dioxide.

5

<hr>

Imaging Fundamentals
and Mathematics

5.1. Purpose of Imaging

Medical images are used to obtain information about internal body organs or the skeleton to determine a patient's physical state. An image may show damage to organs that cannot be externally visualised. The dangers or stress to a patient by use of a non-invasive imaging technique may be greatly less than those of an invasive operation. The latter may also be slow and disturb the physical state which it is intended to evaluate.

A variety of imaging modalities is used. They use variously photon, ultrasonic, thermal and electrical impedance methods to obtain characteristics of the body being examined. These characteristics are evaluated and transformed into a form of picture which indirectly yields information about that patient's pathology.

In principle an image is a representation of some object. Its nature is indirect, and its form is different from the original incarnation. Generally the images we use in medical applications are represented in two dimensions. From examining this representation we may evaluate certain of the characteristics shown in the image to see how they are derived or to see how they deviate from the normal.

Images are thus used to examine unusual growth, which may be cancerous, to examine bone structures which may have been physically damaged, or to look for damage to blood vessels or brain tissue. The visual representation which the image affords may enable a diagnosis of the patient's condition and enable its correction.

This chapter provides a brief overview of imaging technology which is treated more thoroughly in Chapter 6. Our main purpose for now is to introduce the common mathematical background used. This includes a basic formulation of image acquisition which leads on to the methods used to restore and enhance images. For a more thorough treatment of the mathematical background to the signal analysis and image processing covered in this chapter, the reader is referred to specialist texts such as Bracewell (1986) and Gonzalez (1992).

5.1.1. Overview of Imaging Modalities

The various imaging technologies have very different physical characteristics. Sound waves are transmitted reasonably well through tissue. Their velocity is around 1500 ms^{-1}. At this velocity it is reasonable to be able to separate structures within a body by their differential depths. Echoes return in times of up to about 0.1 ms. The hazard from ultrasound energy is believed to be very low at the energies normally used in clinical diagnosis. The resolution of ultrasound is primarily limited by the aperture of the ultrasonic probe, diffraction and thus the wavelength of the sound employed.

On the other hand various high energy photon beams may be used, such as X radiation. These are of an energy which does not permit their focusing, and the velocities are such that pulse echo techniques are not feasible. Unlike ultrasound, however, X rays are transmitted through gaseous spaces, such as in the lungs, and may be used for general thoracic examination. The image formed by X rays is a result of the differential absorption of the radiation in different tissues. The effects of diffraction do not directly limit the resolution of X ray imaging.

As an alternative to using ionising radiation in this fashion, a patient may be given a small amount of radioactive material bonded to a biochemical reagent. 'Nuclear Medicine' studies observe the external radiation produced to help examine, amongst other things, the manner in which the reagent is transferred about the body. This can provide useful diagnostic information of a rather different nature.

X radiation is in principle dangerous. It is generally accepted that there is a risk involved with any exposure to radiation. The risk is normally considered as a stochastic risk with the levels of dose used in imaging. These are likely to cause a number of deaths in a significant population of individuals each of whom receive the exposure. Therefore before any image is obtained, it should be established that the benefit to the patient outweighs the risk of damage due to the examination. We looked at the absorption of radiation in Chapter 3: its safety is further considered in Chapter 8.

Finally, we can also look at the use of NMR scanners. Used in a medical environment, they are normally known as Magnetic Resonance Imaging (MRI) scanners, and obtain a map of the distribution of hydrogen nuclei to provide yet another viewpoint. Other physical parameters are examined in various other imaging systems which are outside the scope of this book. Please refer to the Bibliography for details of literature which may help you.

5.2. Mathematical Background

The mathematical treatment in the following sections is largely based on two dimensional models. This should not be surprising as the form of image we are dealing with is a two dimensional entity. The objects of concern of course are solid (three dimensional) structures, but in the main, the forms of imaging used in medicine obtain two dimensional views.

5.2.1. Linearity

The imaging systems we will discuss will to the first approximation be assumed to be linear. A linear system is one which obeys the superposition principle:

$$S\{aI_1(x,y) + bI_2(x,y)\} = aS\{I_1(x,y)\} + bS\{I_2(x,y)\} \qquad (1)$$

where S is the system function, and the Is are the input functions which are shown in a two dimensional form.

The response of this sort of system may be analysed in terms of its impulse response. The impulse or Dirac Delta function is defined to have a volume of unity:

$$\int\int_{-\infty}^{\infty} \delta(x,y)\mathrm{d}x\mathrm{d}y = 1 \qquad (2)$$

This function is defined to have a vanishingly small width so that its value is zero at all points except (x,y). The function is used in the analysis of systems owing to two important properties. Firstly, it may be used for selection, and secondly as it contains equal power at all frequencies it may be used to characterise a system's frequency response. For mathematical convenience, its shape is defined as that of a Gaussian probability distribution function whose form is shown in Figure 5.1. For a further discussion of this function, please refer to Bracewell (1986).

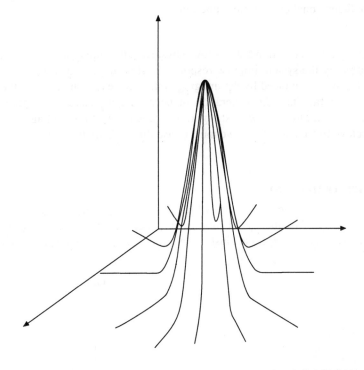

Figure 5.1 Shape of the two dimensional Gaussian Probability function.

The output at a point from a two dimensional system is given by

$$g_2(x_2,y_2) = S\{g_1(x_1,y_1)\} \qquad (3)$$

in which g_1 is the input, and S is the system transfer function which will be considered in more detail later. We use the delta function's selection property over the system in an operation which reorganises equation 3 to obtain

$$g_2(x_2, y_2) = \int\int_{-\infty}^{\infty} g_1(\xi, \eta) S\{\delta(x_1 - \xi, y_1 - \eta)\} d\xi d\eta \qquad (4)$$

where ξ and η are dummy variables of integration. Within the integral in (4) is the system response to a two dimensional delta function. This may be more simply expressed in a different notational form as

$$S\{\delta(x_1 - \xi, y_1 - \eta)\} = h(x_2 - \xi, y_2 - \eta) = h(x_2, y_2; \xi, \eta) \qquad (5)$$

or the point spread function. This assumes that the system response is space invariant, or in other words that S is independent of the values of (x, y). We may now rewrite the system equation as

$$g_2(x_2, y_2) = \int\int_{-\infty}^{\infty} g_1(\xi, \eta) h(x_2, y_2; \xi, \eta) d\xi d\eta \qquad (6)$$

which is a two dimensional convolution function.

$$g_2 = g_1 * h \qquad (7)$$

We consider convolution in section 5.2.5 below. Physically, this equation tells us that the input function is blurred by the system impulse response h. This suggests that we may be able to remove the distortions introduced by the imaging system if we can characterise the system's impulse response to facilitate the generation of better quality images. Unfortunately, the problem will not turn out to be quite so straightforward when we look at image restoration in section 5.3.2 below. Before doing that we must revise the mathematics to be used and clarify its terminology.

5.2.2. Fourier Transform

The Fourier Transform enables a function which is defined in terms of time or space to be specified as a function of frequency. In simpler language, for an image this means that the picture which we see may be expressed instead in terms of a spatial frequency function. As is

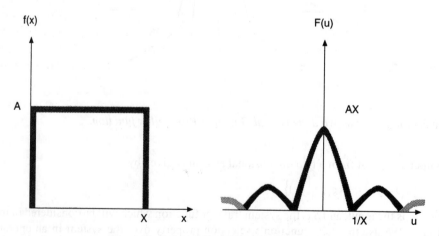

Figure 5.2 Fourier Transform Pair

the case with processing time varying electronic signals, the transformed frequency function is the spectrum of the image in space. The high frequency components therefore represent the fine detail, whereas the lower frequency components contribute the overall structure. The Fourier Transform is given by

$$G(u) = \Im\{(x)\} = \int_{-\infty}^{\infty} g(x)e^{-2\pi i u x} dx$$

(8)

Note that we use the symbol i to represent $\sqrt{-1}$ throughout this book.

For example, consider the function rect (X), which is of value A for $X \geq x \geq 0$ and 0 otherwise. Its Fourier Transform is

$$F(u) = \int_{-\infty}^{\infty} f(x)e^{-2\pi i u x} dx$$

$$= \int_{0}^{X} Ae^{-2\pi i u x} dx$$

$$= -\frac{A}{2\pi i u}\left(e^{-2\pi i u X} - 1\right) = \frac{A}{2\pi i u}\left(e^{\pi i u X} - e^{-\pi i u X}\right)e^{-\pi i u X}$$

$$= \frac{A}{\pi u}\sin(\pi u X)e^{-\pi i u X}$$

A transformed function may revert to its original spatial form by means of the inverse transform

$$g(x) = \Im^{-1}\{G(u)\} = \int_{-\infty}^{\infty} G(u)e^{2\pi i u x} du$$

(9)

The symbols $\Im$ and $\Im^{-1}$ are defined as the Fourier Transform and Inverse Transform operators respectively.

Fourier Transforms of functions exist only in certain circumstances. Fortunately the restrictions are not too severe in the cases in which we are interested. The function $g(x)$ may be transformed if is integrable over its whole domain, it is not infinitely discontinuous and it has a finite number of discontinuities. This statement of the conditions for the existence of the transform is rather formal: it is better developed in specialist texts, such as Bracewell to which the interested reader is referred. If a function can be transformed into the frequency domain, its inverse transform must exist.

The definition of the Fourier Transform pair may be extended to encompass functions of two dimensions:

$$G(u, v) = \Im\{(x, y)\} = \int\int_{-\infty}^{\infty} g(x, y)e^{-2\pi i(ux + vy)} dx dy$$

(10)

and the inverse transform is

$$g(x, y) = \Im^{-1}\{G(u, v)\} = \int\int_{-\infty}^{\infty} G(u, v)e^{2\pi i(ux + vy)} du dv$$

(11)

These are unsurprisingly the forms which we will most frequently encounter when processing images represented in two dimensions.

5.2.3. Discrete Fourier Transform

When we wish to process images or other signals by digital computer, we encounter problems relating to the nature of their storage which is obviously discrete rather than continuous in nature. This leads to two problems. The first is that we have to sample the image data, and the second is that the sampled data becomes a function of a discrete rather than continuous variable. This second problem leads to quantisation noise.

We sample a continuous function $f(x)$ N times at discrete intervals to form a sequence

$$\{f(x_0), f(x_0 + \Delta x), f(x_0 + 2\Delta x), ..., f(x_0 + [N-1]\Delta x)\} \tag{12}$$

which may be considered as the function

$$f(x) = f(x_0 + x\Delta x) \tag{13}$$

in which x has the values $0, 1, 2,...,N–1$ and Δx is the sampling interval.

The sampling process is often described by a sampling function, denoted by the *shah* symbol **III**, which is defined as

$$\text{III}(x) = \sum_{n=-\infty}^{\infty} \delta(x - n) \tag{14}$$

Using this notation the function $f(x)$ is sampled:

$$\text{III}(x)f(x) = \sum_{n=-\infty}^{\infty} f(n)\delta(x - n) \tag{15}$$

When sampling, the sampling frequency must be at least twice the maximum signal frequency. This is described by Nyquist's Theorem. If this condition is not correctly observed, the periodicity of the discrete Fourier Transform (equations 16, 17 below) leads to incorrect distribution of power in the calculated spectrum, causing *aliasing*.

The discrete Fourier Transform is then

$$F(u) = N^{-1}\sum_{x=0}^{N-1} f(x)e^{-2\pi i\left(\frac{u}{N}\right)x}, \text{ for values of } u = 0, 1, 2,..., N-1 \tag{16}$$

and the inverse transform is

$$f(x) = \sum_{u=0}^{N-1} F(u)e^{2\pi i\left(\frac{u}{N}\right)x}, \text{ for values of } x = 0, 1, 2,..., N-1 \tag{17}$$

In these expressions, u/N is analogous to frequency measured in cycles per sampling interval (see for example Bracewell).

Again, this transform pair may be expressed in two dimensions:

$$F(u,v) = (MN)^{-1} \sum_{x=0}^{M-1} \sum_{y=0}^{N-1} f(x,y) e^{-2\pi i \left(\frac{ux}{M} + \frac{vy}{N} \right)}, \tag{18}$$

for values of $u = 0, 1, 2,..., M - 1$, and $v = 0, 1, 2,..., N - 1$ and

$$f(x,y) = \sum_{u=0}^{M-1} \sum_{v=0}^{N-1} F(u,v) e^{2\pi i \left(\frac{ux}{M} + \frac{vy}{N} \right)}, \tag{19}$$

for values of $x = 0, 1, 2,..., M - 1$, and $y = 0, 1, 2,..., N - 1$.

In the case of images, these transforms are often quoted as pairs in a square form, so that $N=M$ and they become symmetric. Equations 18 and 19 may then be rewritten:

$$F(u,v) = N^{-1} \sum_{x=0}^{N-1} \sum_{y=0}^{N-1} f(x,y) e^{-2\pi i \frac{(ux+vy)}{N}} \tag{20}$$

$$f(x,y) = N^{-1} \sum_{u=0}^{N-1} \sum_{v=0}^{N-1} F(u,v) e^{2\pi i \frac{(ux+vy)}{N}} \tag{21}$$

The placement of the constant multiplicative factors N^{-1} is arbitrary.

5.2.4. Computation

The single dimension formulae, equations 16 and 17 above, require in the order of N^2 complex multiplications and additions for their evaluation. This number may be significantly reduced by avoiding the repetition of calculations. A method of reduction of the number of calculations is shown in very brief outline here to endeavour to standardise on the terminology used in the description of the imaging process. For a derivation of the method, known either as the *Fast Fourier Transform*, or the Cooley-Tukey algorithm, see for instance Bracewell. The Fourier Transform of $f(n)$, given in equation 16 may now be rewritten as

$$F(u) = \sum_{n=0}^{N-1} f(n) W_N^{nu} \quad \text{where } W_N = e^{-\left(\frac{2\pi i}{N} \right)} \tag{22}$$

and a similar expression may be written for the inverse transform. Equation 22 may now be split into even and odd parts:

$$F(u) = \sum_{m=0}^{\left(N/2 \right)-1} f(2m) W_N^{2mu} + \sum_{m=0}^{\left(N/2 \right)-1} f(2m+1) W_N^{(2m+1)u} \tag{23}$$

Note that this approach is applicable if N is a power of 2 when the algorithm provides a significant speed improvement. In practice it may be necessary to extend the size of the transformed matrix to meet this condition.

When we split the transform into these two components we may also reorganise it into two functions f_1 and f_2 representing the even and odd parts respectively. But

$$W_N^2 = W_{N/2}$$

since these factors represent the phase. So

$$F(u) = \sum_{m=0}^{(N/2)-1} f_1(m) W_{N/2}^{mu} + W_N^u \sum_{m=0}^{(N/2)-1} f_2(m) W_{N/2}^{mu} \tag{24}$$

and

$$F\left(u + N/2\right) = F_1(u) - W_N^u F_2(u), \text{ for } u = 0, 1, \ldots, N/2 - 1 \tag{25}$$

Here F_1 and F_2 are the $N/2$ point discrete transforms for the f_1 and f_2 sequences respectively. These last two calculations require $(N/2)^2$ complex multiplications each. The second terms require a further $N/2$ complex multiplications. This method thus reduces the number of required operations by a factor of 2 for large values of N.

This same process may now be continued. If the total number of points is a power of two, then it may be carried on iteratively until one point transforms are calculated. Thus for v points, the calculation may be carried out $v = \log_2 N$ times, and the total number of complex muliplications reduced to $(N/2)\log_2 N$.

This process may be extended to cover transformations in two dimensions.

5.2.5. Convolution

We will encounter Convolution frequently in the processing of images. The convolution operation provides the basis for the description of the formation of an image. A full description of the mathematical background is beyond the scope of this book: for a clear introduction, the reader should look elsewhere, such as Bracewell.

The convolution of two functions of x is defined as

$$h(x) = \int_{-\infty}^{\infty} f(u)g(x-u)du, \tag{26}$$

which is more simply expressed by the notation

$$h(x) = f(x) * g(x) \tag{27}$$

u is a dummy variable of integration. An example of convolution is shown in Figure 5.3.

5.3. Imaging Theory

In this section we look at the basis of the composition of an image. The relationship between an object and an image was previously briefly outlined. This may now be explored in more detail to see how to present the image in such a way as to reduce noise, highlight desired aspects and to remove any artefacts introduced by the imaging system. This is again presented in theoretical terms without direct reference to any specific imaging modality: the technology is described in Chapter 6.

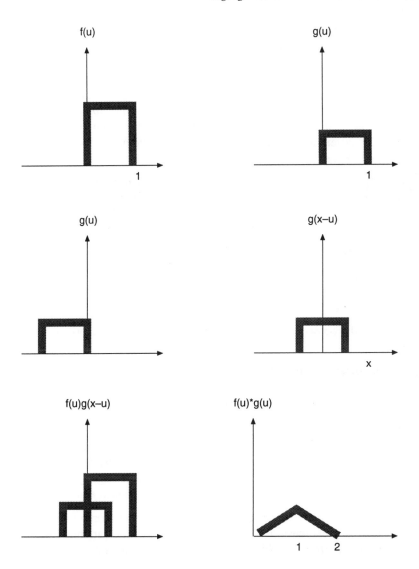

Figure 5.3 Example of convolution

5.3.1. System Transfer Function

An image provides us with a representation of an object. Let the object in space be f and its image g. The image is created by the imaging system as a representation of the object. If the transformation is perfect, then $f=g$. However, real imaging systems never achieve perfection, so we are concerned with the changes introduced. These are in the simplest cases two dimensional representations of solid objects in the manner of a photograph of a solid object.

The simplified general relationship between the object and image is

$$g(x,y) = h(x,y) * f(x,y) \tag{28}$$

This is the convolution of the object f and a function of the imaging system, h. This function in this case is taken to be space invariant, or independent of the values of x and y. This may be expressed in the frequency domain as

$$G(u,v) = H(u,v)F(u,v) \tag{29}$$

where H is called the transfer function. In optical terminology H is called the *optical transfer function*, whose magnitude is called the *modulation transfer function*. Rather in the manner of an amplifier of an electronic signal, the transfer function defines the frequency response of the system. For the system to transfer the finest details of the object to the image, the transfer function should have sufficient bandwidth not to cause deterioration.

When we are going to use computer enhancements of images, it is necessary to sample and digitise the image. For a discrete (or sampled) image, this relationship may be written as

$$g_{i,j} = \iint h_{i,j}(x,y)f(x,y)\mathrm{d}x\mathrm{d}y \tag{30}$$

If we also assume that the object was itself digital, then we may write

$$g_{i,j} = \sum_{k=1}^{N}\sum_{l=1}^{N} h_{i,j,k,l}f_{i,j} \tag{31}$$

which indicates the potential amount of calculation required to restore an image in the real space domain.

Now consider equation 29 in a discrete form, and rearrange it as the inverse transform.

$$f(x,y) = \left(\frac{1}{N}\right)\sum_{u=0}^{N-1}\sum_{v=0}^{N-1}[G(u,v)/H(u,v)]e^{2\pi i(ux+vy)/N} \tag{32}$$

We have attempted in this equation to recover the object f from the image G and the system transfer function H. It should be clear that equation 32 indicates the danger of attempting directly to recover the best interpretation of the object from the image by simple deconvolution. We can clarify this point after introducing a noise term in the next equation.

The system noise may be represented by introducing a further term n into the description of the image and object relationship. The noise term describes the variability between different images obtained from the same object using the same imaging system.

$$g(x,y) = \iint h(x,y)f(x,y)\mathrm{d}x\mathrm{d}y + n(x,y) \tag{33}$$

This equation does not have a unique solution. Rather, we may only achieve a best fit relationship between image and object. If we apply the same technique as in equation 32 to obtain a solution, there would probably be regions of H which are close to zero. These could easily coincide with regions in which the numerator was non-zero, allowing the 'improved' image to become dominated by noise.

We now express equation 33 in rather different terms. Let us consider instead that the f term is an ideal image. Then the degraded image

$$g(x,y) = H[f(x,y)] + n(x,y) \tag{34}$$

This equation rather more clearly describes the operation of obtaining an image as a two stage process whereby the ideal image f is processed by the degradation operation H and subsequently by the additive noise term n to produce a degraded image g.

In our following analysis, we are going to assume that the degradation process is linear, homogeneous and space invariant. Then if we simplify by assuming that the noise is zero, recall equation 1:

$$H[k_1 f_1(x, y) + k_2 f_2(x, y)] = k_1 H[f_1(x, y)] + k_2 H[f_2(x, y)] \tag{35}$$

The noise free solution to the system described by equation 34 is space invariant if

$$g(x - \alpha, y - \beta) = H[f(x - \alpha, y - \beta)] \tag{36}$$

for any α and β.

These expressions form the basis of the description of the image recovery process which we describe in the following section.

5.3.2. Image Restoration

Image Restoration is the process of recovering or improving an image that is in some sense degraded. The physical system which obtained the image we have already observed will have a limited bandwidth and thus will reduce the image's fine detail. It will also introduce noise to the image. It is our purpose in the first place to attempt to obtain an image which is the best estimate of the object from which it was obtained. Expressed in slightly different terms, recall the degradation model outlined in the previous section: we will be seeking to obtain from the degraded image a best estimate of the ideal image.

Returning to equation 4, the non-degraded image may be described in terms of its convolution with the two dimensional impulse function:

$$f(x, y) = \int\int\limits_{-\infty}^{\infty} f(\alpha, \beta)\delta(x - \alpha, y - \beta) d\alpha d\beta \tag{37}$$

Using the noiseless description of the system given in equation 34 above we may substitute for f

$$g(x, y) = H[f(x, y)] = H\left[\int\int\limits_{-\infty}^{\infty} f(\alpha, \beta)\delta(x - \alpha, y - \beta) d\alpha d\beta\right] \tag{38}$$

As the system is linear, it is additive and homogeneous, so

$$g(x, y) = \int\int\limits_{-\infty}^{\infty} f(\alpha, \beta) H[\delta(x - \alpha, y - \beta)] d\alpha d\beta \tag{39}$$

We may define the term

$$h(x, \alpha, y, \beta) = H[\delta(x - \alpha, y - \beta)] \tag{40}$$

which is called the system impulse response. This may be substituted in equation 38 to form

$$g(x, y) = \int\int\limits_{-\infty}^{\infty} f(\alpha, \beta) h(x, \alpha, y, \beta) d\alpha d\beta \tag{41}$$

Equation 41 is known as the Fredholm Integral which states that if the system impulse response is known, then the response to any input $f(\alpha,\beta)$ may be calculated. Alternatively, this says that the system impulse response completely characterises a system.

If the system is space invariant, then equation 41 reduces to the convolution integral:

$$g(x,y) = \int\int_{-\infty}^{\infty} f(\alpha,\beta)h(x-\alpha,y-\beta)\,d\alpha d\beta \tag{42}$$

In a discrete form, equation 42 may be written as

$$g(x,y) = \sum_{m=0}^{M-1}\sum_{n=0}^{N-1} f(m,n)h(x-m,y-n) \tag{43}$$

This may be expressed more compactly in a matrix representation in which **H** is the degredation operator introduced in equation 34

$$\mathbf{g} = \mathbf{Hf} \tag{44}$$

where **g** and **f** are column vectors formed by stacking the rows of the $m \times n$ functions $g(x,y)$, $f(x,y)$. **H** is a matrix of dimension $MN \times MN$.

We now reintroduce the noise term **n** from equation 33 to the equation, again in a vector form (of the same dimension as **g**):

$$\mathbf{g} = \mathbf{Hf} + \mathbf{n}$$

and the best estimate image $\hat{\mathbf{f}}$ is defined by multiplying by the inverse operator $\mathbf{H}^{-1}$

$$\hat{\mathbf{f}} = \mathbf{H}^{-1}\mathbf{g} = \mathbf{H}^{-1}\mathbf{Hf} + \mathbf{H}^{-1}\mathbf{n} \tag{45}$$

or more simply, the best estimate image depends on the idealised image **f**, and the noise term **n** which may have been multiplied by components of the inverse system transfer function $\mathbf{H}^{-1}$, so

$$\hat{\mathbf{f}} = \mathbf{f} + \mathbf{H}^{-1}\mathbf{n} \tag{46}$$

We should also note here that the matrix $\mathbf{H}^{-1}$ may not exist.

Clearly another way has to be found to recover the image other than by direct deconvolution.

In the first place it is rational to try to minimise the effect of noise by using the standard method of least squares. Then

$$\|\mathbf{n}\|^2 = \left\|\mathbf{g} - \mathbf{H}\hat{\mathbf{f}}\right\|^2 \tag{47}$$

which is defined as

$$\|\mathbf{n}\|^2 = \mathbf{n}^T\mathbf{n}, \text{ and } \left\|\mathbf{g} - \mathbf{H}\hat{\mathbf{f}}\right\|^2 = \left(\mathbf{g} - \mathbf{H}\hat{\mathbf{f}}\right)^T\left(\mathbf{g} - \mathbf{H}\hat{\mathbf{f}}\right) \tag{48}$$

where the T operator transposes the matrix.

Now minimise equation 47 by setting its differential to zero:

$$0 = 2\mathbf{H}^T\left(\mathbf{g} - \mathbf{H}\hat{\mathbf{f}}\right) \tag{49}$$

or

$$\hat{\mathbf{f}} = \left(\mathbf{H}^T\mathbf{H}\right)^{-1}\mathbf{H}^T\mathbf{g} \tag{50}$$

$$\hat{\mathbf{f}} = \mathbf{H}^{-1}\mathbf{g} \tag{51}$$

Which is unfortunately the result previously obtained in equation 32 that caused noise amplification and was therefore of no use.

This problem may be resolved by using a modified technique, which instead of attempting to recover the image itself, seeks to find a best modified form of the image.

We will seek to minimise a function of the form $\left\|\mathbf{Q}\hat{\mathbf{f}}\right\|^2$ under the constraints imposed in

equation 47. $\mathbf{Q}$ is a linear operator which operates on $\mathbf{f}$. Applying the constraint, we may write a function

$$J\left(\hat{\mathbf{f}}\right) = \left\|\mathbf{Q}\hat{\mathbf{f}}\right\|^2 + \alpha\left(\left\|\mathbf{g} - \mathbf{H}\hat{\mathbf{f}}\right\|^2 - \|\mathbf{n}\|^2\right) \tag{52}$$

In this equation, α is known as a *Lagrange Multiplier*. Differentiating this function $\partial J\left(\hat{\mathbf{f}}\right)\Big/\partial\left(\hat{\mathbf{f}}\right)$ and then minimising by setting the differential to 0 yields

$$\hat{\mathbf{f}} = \left(\mathbf{H}^T\mathbf{H} + \tau\mathbf{Q}^T\mathbf{Q}\right)^{-1}\mathbf{H}^T\mathbf{g} \tag{53}$$

Here we have substituted $\tau = 1/\alpha$ for clarity. Now if τ is set to zero, this equation reverts to the direct deconvolution we examined earlier.

The operator $\mathbf{Q}$ may be selected now to enable values of $\hat{\mathbf{f}}$ to be obtained which are not dominated by noise in parts of the spectrum. This technique is explored in more detail by Gonzalez (1992).

5.4. Image Processing

The following sections outline several forms of 'improvement' which are undertaken to clarify aspects of images. This treatment is far from exhaustive: the interested reader is referred to specialised texts on image processing, such as Gonzalez (1992).

5.4.1. Enhancement

An image is enhanced to draw out certain characteristics that would otherwise be unclear. It is therefore different in principle from restoration, and is involved with the image itself rather than the manner in which it was produced. The sorts of enhancement discussed here involve filtering and contrast enhancement which are used to alter the emphasis on particular image features. In a medical context these improvements are undertaken to improve the diagnostic features of the image.

The methods are therefore not general: the particular form of enhancement that is indicated for a form of problem is likely to be specific to that problem and imaging modality. The major groups of methods employed are those which operate in the Spatial and Frequency domains respectively.

The first forms of spatial enhancement methods are those which operate on points in the image. These transform the grey level of points. This may be desirable for several reasons. The first is that the dynamic range of the image information exceeds that of the display medium. For example, there may be an image with components distributed over a range of 10^6, to be shown on an 8 bit display. This form of image may be displayed with greater clarity by logarithmic compression of its dynamic range. On the other hand, an image may show its features better if much of the grey scale information is hidden by altering the display characteristics so that it displays only a limited dynamic range, perhaps only two bits. The threshold level for this transition needs to be selected to give the best view of the desired features.

A more general form of point image manipulation is to use a histogram method. Here the starting point is to obtain the distribution densities in the image of each discrete grey level. Transformations may be undertaken either to shift the overall distribution or to modify the image brightness. Alternatively the density distribution function may be expanded or contracted in order to modify the resulting image contrast.

Techniques are also employed to modify the whole image in order to equalise the histogram of its grey scale, to make it fit a defined pattern, or to modify selectively different areas of an image according to different recipes. These all have characteristics which lend themselves to particular situations.

The other aspect of spatial processing of concern involves undertaking transforms based on the properties of an area rather than single points individually. These processing techniques use filters which may be used either to smooth out transitions (low pass filters) or to enhance contrast (high pass filters). If the filters are matrices populated by numbers used to multiply the points in the image's matrix, and they are defined symmetrically about their centre, they in effect carry out the convolution operation directly. These filters are linear in operation.

Other spatial filters may be defined which are not linear. An example is the median filter. This examines points in a neighbourhood, and sets their values to those of the median value in the set of those points. For example the filter would select a group of nine points (a 3 × 3 array), and set their values in ascending order. The value found for the middle element – the fifth – is used to replace the value of the pixel. This technique quickly removes high noise peaks in an image as they have values which are excluded from being the median value. It has the distinct advantage over smoothing filters, which carry out a low-pass filtering operation, that it does not directly affect edge sharpness and thus lose detail.

Filtering may also be carried out in the frequency domain. These may be defined either as ideal filters, or to have a particular cut off shape. Frequency domain filtering is attractive either when it is desired to match a spectrum exactly, or when the image data are already held in the frequency domain for other reasons.

5.4.2. Resolution

The resolution of an image is defined as the ability to enable two points to be distinguished. It is therefore specified as the maximum spatial frequency in the object which may be recognised in the image. This is clearly a function of frequency and contrast.

The primary factor which influences the resolution of X ray images is therefore the sampling frequency. For tomographic scanners this is due to the number of samples taken when rotating the imaging system around the patient. In gamma camera or other scintillation counter systems, the size and number of sampling points defines the resolution. For a formal definition of the term 'resolution', see for example Longhurst (1967).

Thus for a tomographic image, the resolution is traded against the total X ray dose given to the patient. The more samples taken, the greater the overall dose given, but the better the spatial frequency information.

5.4.3. Blur Removal

It is possible to remove some of the movement artefacts from digitally stored images which are likely when the image sampling time extends beyond a period of about a tenth of a second. It is likely that many medical images suffer from such artefacts: certain of the more modern imaging equipment is able to assist in their removal.

5.4.4. Image Analysis

A further topic to be considered is image analysis. The term relates to the examination of the image to identify its structural components. The image may be segmented, by examining it to locate edges and thus other geometric shapes. These are the building blocks used subsequently for recognition of the structures in images. This morphological analysis is able to detect structures such as text characters in images from an assemblage of pictorial regions.

The analysis of regions requires the application of Knowledge Based System techniques. Image structures require to be compared with known structural forms. In a medical application, we should know which structure we are looking for, but it may be desirable to use an automatic system to attempt to quantify some aspect of the information stored in an image. In many areas of image analysis the objective is to make a ready and quick assessment of a pattern. In medical analysis, rather the reverse is the case, as it may be desirable to isolate the unrecognised or abnormal so that it may be enhanced and more clearly displayed.

A serious limitation remains that the recognition of features is difficult when applied to noisy images. Improving the signal to noise ratio of medical images is difficult. Increased exposure is required to obtain an improvement potentially leading to the use of greater doses than would otherwise be required. This brings in the ethical question about when to undertake radiological examination (see Chapter 8). Lengthening the duration of exposure is likely to increase movement artefacts potentially obliterating the improvement obtained in the physical image signal to noise ratio.

Some recent imaging equipment uses these techniques to recognise structures, such as the chambers of the heart so that the stroke volume may be measured by imaging technology rather than using an invasive technique.

6

Imaging Technology

The purpose of this chapter is to introduce a number of the currently used methods of obtaining medical images. The list is far from exhaustive, but is intended to provide a basis of the techniques used.

We intend to show from the range of these techniques that although none is a panacea, that each provides characteristic information which can assist a clinician in certain types of diagnosis. Examination for certain conditions may well require the application of more than one method. Sometimes it may be of use to attempt to construct single images derived from a combination of different modalities in order to build up a clearly recognisable picture of a particular pathology.

All the techniques outlined in this chapter are progressively being refined by the computer systems which are integrated with them. Manufacturers are now offering both standard network interfaces to enable remote access to the image data and optical disc storage to permit cost effective long term storage of digital images.

6.1. Projection X Radiography

The simplest, earliest developed and most frequently used form of medical imaging is by the use of projection X radiography. X rays are shone through the area of the patient under study, taking care to avoid exposure beyond the bounds of interest and to use the energy of X radiation which best shows the aspects of the patient being studied. In the simplest routine forms of processing, a latent image is formed on an X ray film which is subsequently developed and fixed to make the image visible. We outline the technique in Section 6.1.1 together with its technological descendant which uses digital image acquisition to enable images to be processed.

Projection X radiographs are either made directly by effectively comparing the absorption of tissues, or may be enhanced by the administration of a radiopaque dye which is localised in either tissues or fluids. Plate 1 (p. 133) shows the sort of apparatus used to obtain these images.

6.1.1. Film Characteristics

An ordinary photographic film which is sensitive to visible light is made from a series of layers coated onto a cellulose medium. On the active side of the film there is, from the outside in, a protective coating, a light sensitive emulsion, and a layer to assist the emulsion to adhere to the cellulose. The reverse of the cellulose has a backing layer whose purpose is to reduce curling. The film emulsion is made of a colloidal suspension of silver halide crystals in a gelatine medium. These are of an extremely small particulate size. Each grain may individually be blackened as a result of the absorption of light as energised grains may be chemically reduced by the subsequent action of film developer to silver. Any grains which are not chemically reduced are purged of their silver halide by the action of a fixing agent which thus prevents further blackening. (Note that a reverse or positive image may be directly obtained by instead purging the film of silver, and then reducing all remaining silver halide to silver.) The action of light on silver halide is to move electrons into trapping sites in the crystal from which they may be removed by the reducing action of the developer.

The sensitivity of this sort of film to light is largely controlled by the grain size. Larger grains are more likely to have been sufficiently activated to enable them to be reduced, so the film is more sensitive to light, but has a poorer resolution as a result. These faster films also have a shorter region between being fully exposed and clear: they therefore have a higher contrast.

A film has the characteristic typically where there is a '*fog level*', or maximum transparency which sets the lower limit of its sensitivity. Its blackening response to increasing light increases logarithmically to a shoulder, beyond which it is saturated. The slope of the logarithmic region is the *contrast* of the film. These characteristics are shown in Figure 6.1.

An X ray film differs from the description above because of the limited absorption probability of X ray photons by the silver halide emulsion. Unlike a visible film, X ray cellulose film bases are coated on both sides with the silver halide emulsion. This increases sensitivity at the price of some loss of film resolution. However, the system is not limited by film resolution,

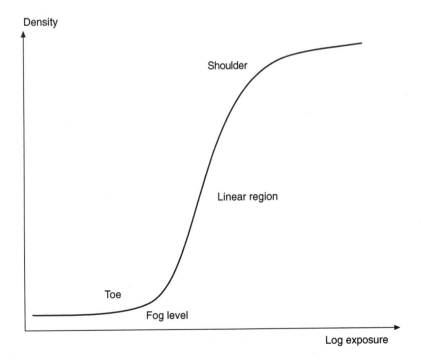

Figure 6.1 Photographic film characteristics

and it is desirable to maximise film sensitivity, as it is not usual to enlarge shadow X radiographs significantly. The resolution of the film is in fact limited by the mean free path of the secondary particles produced as a result of the incidence of X rays on the film. These have mean free path lengths of between 10 and 100 μm, whereas the thickness of the film emulsion is around 20 μm and the base thickness about 200 μm. The emulsion thickness therefore effectively limits the spread of the secondary radiation, so direct emulsions have good resolution characteristics where a relatively high dose may be tolerated.

As an alternative to these direct emulsion films, there are films which are used in conjunction with fluorescent screens. The phosphor is made of particles of sizes of about 10 μm diameter in a layer between 70 and 300 μm thickness. X ray photons incident on the phosphor produce secondary visible photons, but these are released in a variety of directions causing a loss of resolution. There are configurations in which either one or two phosphors are used to assist in dose reduction: there is a trade off between dose and resolution.

6.1.2. Dose Considerations

The required dose in projection X radiography where film is used as the receptor is controlled by the sensitivity of the film and the need to ensure that an adequate signal is available to obtain the required signal to noise ratio. The X ray spectrum used is chosen to obtain satisfactory differential absorption in the range of different tissues encountered for a particular image type.

It will be seen below that the spectrum incident on the patient is dependent on the tube excitation potential, the target (anode) material and the form of filtering used between the tube and patient. In broad terms low energy radiation is more readily absorbed than higher energies. Therefore when looking through greater depths of tissue, the use of higher energy beams is indicated. The use of low energy components is simply damaging as their energy is released to the tissue without contributing to forming an image.

The power for an X ray tube is best derived from a stabilised power supply. This means that the spectrum produced is controlled. The energy spectrum of the tube is characterised by a peak emission determined by the instantaneous excitation voltage, and a low end defined by absorption of the radiation by the material through which it passes. This is normally controlled by the use of filters which absorb components below a certain energy. There are normally other significant peaks in energy emission due to characteristic radiation of k shell electrons from the target material.

The radiation dose received by a patient during this form of radiography is typically between 1 and 10 mGy. The actual dose delivered is very variable, as it depends on the nature of the examination undertaken, and therefore on the type of equipment employed. Perhaps more surprisingly the dose for a given type of examination appears to vary between different hospitals undertaking the same examination. The guidance to the Medical Physicist should be to take whatever steps are necessary to minimise the dose received to a level compatible with an acceptable level of image noise. Doses should be consistently 'As Low as Reasonably Achievable' (ALARA) when proper quality control measures are in place and the radiological apparatus is correctly set up. (See, for instance, Carmichael, 1988).

6.1.3. Attenuation of Beam by Tissue

The photons in an X ray beam passed through a patient are either transmitted, scattered or absorbed. The removal of photons from the beam by scattering and absorption may be expressed as

$$\Delta N = -\mu N \Delta x \tag{1}$$

where ΔN represents the number of photons lost from a population of N incident photons over a distance Δx. The absorption coefficient is μ. We may integrate and solve equation 1 to obtain the loss through a body of thickness x:

$$N_{out} = N_{in} e^{-\mu x} \tag{2}$$

Instead of looking at the number of photons directly, we may instead use the intensity of the beam and expand the absorption coefficient to take account of its spatial variation and dependence on incident photon energy. Equation 2 becomes

$$I(x, y) = \int I_0(\varepsilon) e^{-\int \mu(x, y, z, \varepsilon) dz} d\varepsilon \tag{3}$$

in which $I_0(\varepsilon)$ is the incident intensity, and is a function of the photon energy ε.

These expressions represent the total absorption of the beam. Photons may release their energy in either a single collision or a series of events. Now we may consider the mechanisms which give rise to the removal of the photons. This treatment builds on the material we presented in Chapter 3 which describes the forms of collision process undertaken by ionising radiation.

Firstly, photons are removed by Rayleigh Scattering. The wavelength of lower energy components of a diagnostic X ray beam is in the same order as the atomic diameter. As a result these components may be diffracted at effectively random angles owing to the lack of structural coherence in tissue.

At higher energies, the photoelectric effect predominates. Photons incident with tissue may give up their energy to tightly bound electrons which recoil in the tissue. They are in turn displaced by neighbouring electrons which release fluorescent radiation. This effect varies as the third power of the atomic number. There are significant peaks in absorption which depend on the ionisation energies of each electron band: the most pronounced is the k shell absorption edge. This energy is below the region of interest for most biological materials, although it will be seen to be significant in the spectrum of X rays produced by diagnostic sets.

The third scattering mechanism which is of concern is Compton Scattering. This is the effect whereby a free electron receives kinetic energy from an incident photon and is displaced. The photon therefore loses energy, but in the region of interest, only a small percentage. The scattered photons may thus be detected as coming from a different direction. This effect must be countered to avoid loss of image contrast.

Figure 6.2 shows the relative contributions of each of these absorption mechanisms in water as a function of incident photon energy.

Differentiation of the various forms of bodily tissues is normally the result of the differing absorption coefficients for those materials. The proportion of calcium with its higher atomic weight accounts for the larger absorption coefficient for bone than other forms of tissue. The

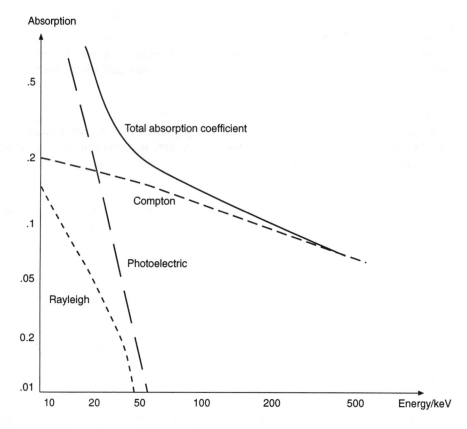

Figure 6.2 Absorption coefficient as function of energy

coefficients tend towards common values at higher energy levels once Compton Scattering predominates. These are plotted in Figure 6.3.

6.1.4. Scatter and Image Contrast

X ray image contrast is affected by scatter. Photons passing through a patient are transmitted, absorbed and scattered (as described in section 6.1.3). The receptor is sensitive to those photons which are not absorbed. Reducing the number of received scattered photons may be partially achieved by collimating the X ray beam immediately in front of the receptor to constrain the range of angles of incidence of photons which will be counted by the receptor.

Before proceeding, we define the ideal image contrast. This is, in the absence of noise,

$$C = \frac{\Delta I}{\bar{I}} \tag{4}$$

where $\bar{I}$ is the mean background intensity and ΔI the local intensity variation.

With X rays the scattering process is caused by the interaction of photons with atoms. The scatter probability is a function of both the type of material and the photon energy. Generally higher energy photons are less likely to be scattered, and low energy photons are more likely

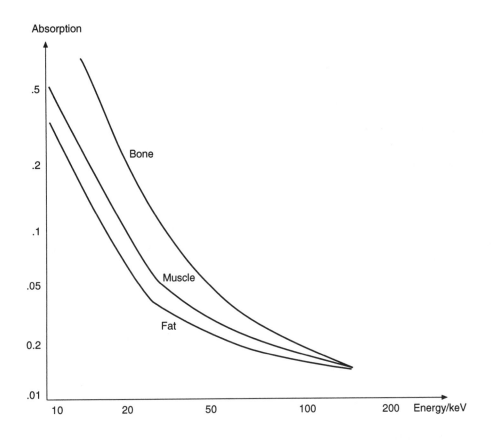

Figure 6.3 Absorption coefficients for tissue as function of energy

to be absorbed completely, either by the photoelectric effect or multiple scattering. The effective X ray dose is normally higher as a result of the use of low energy X rays than high energy beams.

Two measures may be taken to mitigate the effects of scatter. Firstly, it is normal practice to place a collimating screen in front of a film. When digital radiology is performed, as described in section 6.1.6, with certain forms of image receptor it is usual to scan the object with X rays. This means that a properly collimated beam may be recorded with the effects of scatter considerably reduced.

Another major restraint on image quality is due to source noise. In the case of a communications system, the signal frequencies typically range up to around 1GHz. With X ray systems, we are measuring photons which are released from their source by quantum mechanical processes. Their energies therefore spread with a distribution of energies defined by a Poisson distribution. The thermal noise in a system is shown by Bleaney (1976) to be $4kTB$, where k is Boltzmann's constant, T absolute temperature, and B the bandwidth of measurement. The noise energy time $1/B$ is therefore $4kT$. The energy carried by a photon is $h\nu$, so for a population of N photons, the signal to noise ratio is

$$\text{SNR} = \frac{Nh\nu}{\sqrt{Nh\nu + 4kT}} \tag{5}$$

The term $\sqrt{N}h\nu$ is the standard deviation in photon energy. With X ray frequencies, ν, of around 10^{19} Hz, the ratio $4kT/h\nu$ is 2.5×10^{-6}, so the signal to noise ratio is dominated by quantum noise and is proportional to $\sqrt{N}$.

The effect of this quantum noise is to introduce what is known as an *image mottle* to X ray images. Clearly it is reduced by increasing the X ray dose used to obtain the image.

The other main limitation of image quality to be considered here is that due to scatter. Recalling equation 4, if we take account of the degradation of the image contrast due to scatter, this may be expressed as

$$C_r = \frac{CI_t}{I_t + I_s} = \frac{C}{1 + I_s / I_t} \tag{6}$$

Here, C_r is the resultant contrast, degraded by the effect of the additional intensity I_s of scattered photons to the transmitted flux I_t. In terms of the image signal to noise ratio equation 4 may be expressed (see Macovski, 1983)

$$\text{SNR} = \frac{C\eta N}{\sqrt{\eta N + \eta N_s}} = \frac{C\sqrt{\eta N}}{\sqrt{1 + N_s / N}} \tag{7}$$

where η is the detector efficiency. This contrast loss will occur unless steps can be taken to mitigate its effects. One method is to collimate radiation ahead of the receptor: this is described in section 6.1.5 below. The other is to attempt to discriminate the energy levels of incident radiation, as the Compton scattered radiation has a lower energy than that of source photons. This latter technique is only practicable for monochromatic sources.

6.1.5. Beam Collimation

The output of a diagnostic X ray tube is controlled either by a gate to constrain its spread or by a collimator to produce a parallel beam. As the resultant image contrast and hence its resolution depends significantly on the amount of received scatter, the beam is passed through a collimator, or grid, in front of the receptor.

The grid (shown in Figure 6.4) is made of strips of lead, separated by another material. For application with low energy beams, this is normally cotton, whereas with high energy beams, aluminium is used.

The grid may be left stationary, but will then cause the display of the grid lines. Alternatively it may be moved in some form of reciprocating motion. The grid chosen is dependent on the application. Densities of 80 to 150 lines cm^{-1} are employed. Moving grids with a line density of 30 cm^{-1} and a cotton or paper interspace are used where minimising the dose is crucial.

A further alternative method is to reduce the X ray source to produce a line of radiation which traverses the patient, and to provide a pair of slits above the receptor which track the rotation of the beam of X rays. The slits have to be moved at different velocities, but this arrangement may significantly reduce the effect of scatter on image quality. Its disadvantages are its mechanical complexity and the time taken to scan the area of interest which can lead to differential movement artefacts.

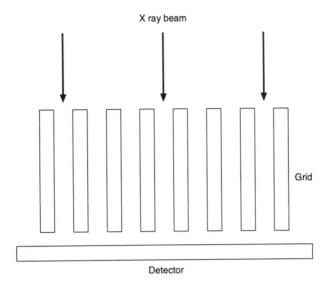

Figure 6.4 Scatter reduction grid

6.1.6. Image Construction

The description given above primarily refers to the use of X radiographs taken with a film receptor. Instead this may be replaced with a real time detector which enables the image to be recorded and then processed by a computer. Clearly this makes possible image enhancement but at a cost of significantly more expensive apparatus. (Image enhancement can be carried out using film as an intermediate recording medium so that enhancement and manipulation are undertaken off line.)

Early forms of fluoroscopic system used a phosphor coated screen in place of the film to enable direct viewing by the radiologist. This represents a radiation hazard for the operator who over a period receives a significant dose. This apparatus has poor performance in terms of its signal to noise ratio if the patient dose is to be kept within reasonable bounds.

More modern real time methods use image intensifiers. These use a fluorescent screen in close contact with a photocathode. This cathode is inside a vacuum tube with an excitation potential of around 25 kV. The system has a series of interfaces.

1. The X ray phosphor has an efficiency of somewhat less than unity, so not all X ray photon energy is converted into visible photons.

2. However, on average each X ray photon absorbed produces an equivalent of around 1000 visible photons.

3. The photocathode has an efficiency of about 10% in conversion of visible photons into electrons.

4. The output phosphor converts each excited electron into about 1000 visible photons.

The system overall produces around 10^5 visible photons for each five received X ray photons. This gain is necessary to compensate for the sensitivity of the eye if observed directly:

alternatively this system lends itself well to being coupled to a video camera so that the image may be recorded, displayed, and if necessary enhanced. The form of an image intensifier used for radiology is shown in Figure 6.5.

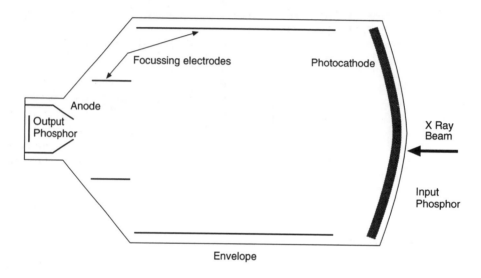

Figure 6.5 X ray image intensifier

As discussed so far, digital image acquisition would offer little more than film. Images could be processed and potentially analysed to obtain some degree of automation of search. Having image information stored in a digital fashion and able to be processed affords several significant advantages.

Firstly, consider an image of 1024 pixels square. If it is recorded at a resolution of 8 bits, then it would require 1MB of storage. Images typically require examination in the immediate aftermath of being obtained, and potentially some further examination in the following period of days. In some cases it is desirable to be able to compare images obtained over a period of months or years to be able to observe the progression of a condition. In any event, there is a common legal requirement to retain images for around a ten year period in common with other recorded medical diagnostic information. This quantity of information is sufficient to cause serious problems with its management and retrieval. Current magnetic disc technology will store around 1–10 GB on a single disc volume, and high density tapes a comparable amount. A hospital may require to store image data amounting to around 1000 GB in a year. Potentially, digital media could reduce the space requirements for this storage as well as providing a more manageable means of access.

The image quality depends also on the uniformity of illumination provided by the X ray source and the spatial response of the detection system. These problems may be addressed by standardising the system spatial response in the image processing system. Furthermore, the use of digital systems permits the acquisition of a series of images over a period of months: the images may be compared with one another having used rubber sheet mapping to obtain the best correspondence between separately obtained images.

From the clinical viewpoint, digital radiology provides superior features. Apart from being able to improve images by filtering, noise reduction and edge enhancement, it makes possible other modes of imaging. The most significant involve taking two images in slightly different ways, but in register so that their differences may be examined. The removal of common information by subtraction means that an image is built which is devoid of the obscuring detail of the anatomy not under examination. Thus although there is an inevitable increase in noise as the images are obtained in the region where quantum mottle is significant, the result is a much clearer view of the desired clinical information.

The major technique employed is known as Digital Subtraction Angiography. Here, an image of the region of interest of the patient is obtained, followed seconds later by taking a further image after the administration and perfusion of a radiopaque dye. Subtracting the first image from the second yields a view of the major blood vessels without the obstructing detail otherwise presented by other anatomical features. An obvious limitation with this technique is that the period between snapshots is sufficient time for movement to occur and thus reduce the quality of the image obtained. Imaging of blood vessels can indicate a number of clinical conditions, such as narrowing or obstruction, aneurysms (the weakening of the wall of a vessel which is a frequent prelude to its rupture) and displacement of the vessel due to a growth. Plate 1 is a picture of an X ray tube and detection system used in digital subtraction angiography and a peripheral image obtained is shown in Plate 2.

As a further alternative to this 'temporal' subtraction, it is possible to discriminate between different tissue types, as we noted in Figure 6.3 above, by the use of different X ray energies. This technique, known as 'energy' subtraction, enables views to be obtained of one form, perhaps all bone, or all soft tissue, affording images perhaps of calcification in blood vessels

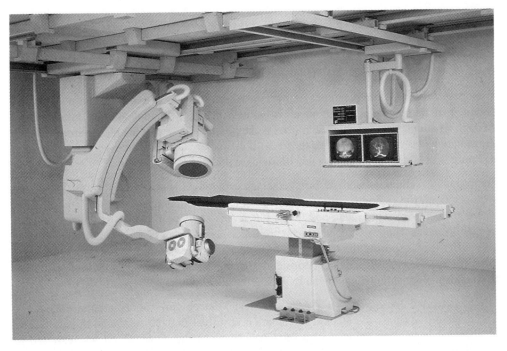

Plate 1 X Ray table used for fluoroscopic examination

which would be too subtle to view without subtraction (see Brody, 1983). The technique employs deriving scanned images in which the scan lines are switched between alternately high and low energy X ray beams. The scan and switching speeds are sufficiently high to minimise any movement artefacts.

These techniques complement the use of tomographic scans, described in section 6.2, as they afford good images of blood vessel structures.

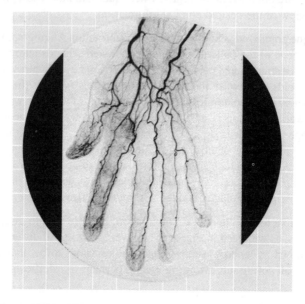

Plate 2 Peripheral image obtained using Digital Subtraction Angiography

6.2. Computerised Tomography

Tomography is the technique of providing a visualisation of a section. It enables a view to be obtained of the spatial distribution of structures including their depth. By taking views through a patient, tomographic techniques make possible a view of the plane through which the radiation was shone. The image then is as if the patient had been cut right through so that a picture is obtained looking perpendicularly onto the cut.

Tomographic techniques were first implemented using moving sources and films. They were swung around the region of interest and thus blurred components other than those about which they rotated. This made possible a much clearer view of certain structures, but the images do not have improved contrast over projection X radiographs. These *Motion Tomographs* also suffer from exposing the whole area of the patient under study to a significant radiation dose. If more than one area is under consideration, the movement prevents multiple examinations taking place simultaneously, so requiring further exposure.

In Computerised Tomography (CT), the source and detector are rotated around the patient. The detector measures the amount of radiation absorbed in each rotational direction. The information is used to reconstruct a sectional image of the plane which was exposed to radiation. Clearly the more sectional images that are obtained the greater the radiation dose.

Doses due to tomographic examination are significantly higher than those of conventional shadow X radiographs, the examination technique is slower and requires expensive apparatus. It is therefore not used as an examination technique unless the additional clinical information indicates its requirement.

6.2.1. Technological Outline

In the earliest generation of apparatus used to obtain computer tomography images, the source and detector were mechanically moved around the patient. Scanning was by stepping them in concert in a single plane across the patient, then rotating through a small angle and repeating the procedure. The source was a finely collimated beam of X rays. This procedure is clearly lengthy: scans to produce a single section took typically four minutes, which clearly requires that the patient be held in a single position if movement artefacts were not to degrade the image seriously.

Subsequent generations of scanners have reduced the requirement for mechanical movement, firstly by employing groups of detectors which received a fan shaped planar beam of X rays. Clearly a technique in which either the beam is spread out, and is potentially non-uniform, or one which relies on a number of different detectors requires that it shall be stabilised to ensure that the system response is uniform.

The current generation of equipment employs a rotating source which emits a fan shaped beam of X rays through the patient towards a circular array of detectors. This arrangement, if there is some overspill of the beam, permits the sensitivity of each detector to be compared with others and hence standardised during the scan. These scanners are capable of completing a section scan in around 2 s. The resolution of the scan is maximised by shifting the diameter of rotation of the X ray source according to the diameter of the section of the patient of interest. Apart from obtaining a scan in a single slice, modern equipment also permits the use of helical scanning: circular and helical scanning methods are contrasted in Plate 3.

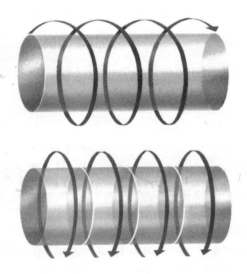

Plate 3 Circular and helical scanning methods

Webb (1990) describes a further scanner which used a static source which worked by being swept about an annular target to produce the X ray beam. This is quoted as being able to complete a scan in better than 0.1 s and thus able to freeze cardiac motion.

6.2.2. Image Restoration

Two main methods exist for the restoration of images obtained by tomographic sectioning. The first is an iterative method, which involves estimating an initial image and then progressively improving it by comparison with the data obtained. The method requires relatively large amounts of memory to hold its control arrays and is not simply described in an analytic fashion. The other is derived from analysis of the imaging mathematics in the frequency domain. We present a feasible method for a frequency domain image restoration technique from tomographic data. This assumes a simple geometry for the scanner which uses a single collimated X ray beam.

Practical scanners require rather different assumptions about their geometry, since, in practice, modern machines use fan beam geometries.

6.2.3. Fourier Methods

Recall that the form of information which is obtained is a series of samples of the absorption of radiation summed through the thickness of the body. They are measured in groups found by stepped movement across the patient. As a first step we consider the form of this projection in purely mathematical terms. The absorption is modelled by a two dimensional function $f(x,y)$, and its projection, for simplicity in the direction of the x-axis, is

$$g(y) = \int f(x, y) dx \tag{8}$$

which is a set of line integrals in the x direction. This projection is shown in Figure 6.6. The Fourier transform of the function $f(x,y)$ is given by

$$F(u, v) = \iint f(x, y) e^{-2\pi i(ux + vy)} dxdy \tag{9}$$

In the direction of spatial frequency $u = 0$, equation 9 may be rewritten

$$F(0, v) = \iint f(x, y) e^{-2\pi i vy} dxdy \tag{10}$$

$$= \int \left[\int f(x, y) dx \right] e^{-2\pi i vy} dy$$

$$= \Im\{g(y)\}$$

which is the one dimensional Fourier Transform of $g(y)$.

This gives us the relationship between the one dimensional Fourier Transform of the projection $g(y)$ and a particular projection of the transform of the two dimensional function $f(x,y)$. This is known as the Central Slice Theorem.

We may generalise the projection of the image by taking transforms for each projection as the imaging system is rotated. In this case the radial projection density function is

$$g_\theta(R) = \iint f(x, y) \delta(x \cos\theta + y \sin\theta - R) dxdy \tag{11}$$

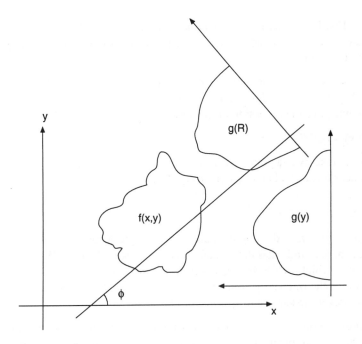

Figure 6.6 Projection of a shape

where the delta function selects the projection of interest at an angle θ from the *x*-axis, and *R* is the line

$$R = x\cos\theta + y\sin\theta$$

As it will be more convenient to model the system in polar co-ordinates, we may write the Fourier Transform of $f(x,y)$ in polar co-ordinates too. The Fourier Transform in equation 9 is then

$$F(u,v) = F(\rho,\beta) \tag{12}$$

$u = \rho\cos\beta$ and $v = \rho\sin\beta$. This substitution may be inserted in equation 9

$$F(\rho,\beta) = \iint f(x,y)e^{-2\pi i\rho(x\cos\beta + y\sin\beta)}\,dxdy \tag{13}$$

Now selecting a section along a line $R = x\cos\beta + y\sin\beta$ using the Delta function yields

$$F(\rho,\beta) = \iiint f(x,y)\delta(x\cos\beta + y\sin\beta - R)e^{-2\pi i\rho R}\,dxdydR \tag{14}$$

This may be simplified using the Central Slice Theorem quoted in equation 11 to give us

$$F(\rho,\beta) = \int g_\beta(R)e^{-2\pi i\rho R}\,dR \tag{15}$$

which is

$$F(\rho,\beta) = \mathfrak{F}\{g_\beta(R)\}$$

Equation 15 relates the Fourier Transform of a two dimensional function to the single dimensional Fourier Transform of its projection. The set of Transforms of projections at angles β may threrefore be used to recreate the polar form of the Fourier Transform of the two dimensional function $f(x,y)$.

We are now in a position to apply this model to the problem of reconstructing a tomographic image which can be obtained as a set of projections. We start from a simplified statement of Beer's Law which is an expression of the absorption of X radiation:

$$I(x) = I_0 e^{-\mu x}$$

where I is the intensity at a point, I_0 the incident intensity and μ the absorption coefficient. So along the linear track of a beam, the intensity is given by:

$$I_\phi(x') = I_{\phi,0}(x')e^{-\int \mu(x,y)dy'} \tag{16}$$

where the beam is directed along the path y' at an angle ϕ, and $I_{\phi,0}$ is the incident intensity. Note that the mark (') indicates the variables associated with the path. We need to calculate the function $\mu(x,y)$ which is the spatial absorption function, and will become our image. Its representation in the frequency domain is (conventionally) M which may be calculated using the theorems developed above from the set of projections.

Note that these expressions assume that all measured radiation has passed along the required direction. This is a simplification in which scatter has been ignored, or eliminated from detected beam and that the beam itself is monochromatic (ie. of a single energy).

Now applying the inverse of equation 14 with a sampling function ρ

$$\mu(x,y) = \int_0^\pi \int_{-\infty}^\infty M(\rho,\phi)e^{2\pi i\rho(x\cos\phi + y\sin\phi)}|\rho|d\rho d\phi \tag{17}$$

Equation 17 may be split into two parts to facilitate its solution.

$$\mu(x,y) = \int_0^\pi \lambda(x')d\phi\Big|_{x'=x\cos\phi + y\sin\phi} \tag{18}$$

where

$$\lambda(x') = \int_{-\infty}^\infty M(\rho,\phi)e^{2\pi i\rho x'}|\rho|d\rho$$

and $\lambda(x')$ is the filtered projection. It is obtained by sampling the absorption of radiation through sections – the evaluation of the integral in equation 16 – and is

$$\lambda_\phi(x') = -\ln(I_\phi / I_{\phi,0})$$

The Fourier Transform of the projection λ at an angle ϕ is

$$\lambda_\phi(x') = \int_{-\infty}^\infty \Lambda_\phi(\varsigma)e^{2\pi i x'\varsigma}d\varsigma \tag{19}$$

The central slice theorem, quoted in equation 15 above, states that

$$\Lambda_\phi(\varsigma) = M(\rho, \phi) \tag{20}$$

The function ρ in equation 17 is the bandwidth limiting or filtering function. It indicates the requirement that there need to be sufficient samples of M at angles ϕ for the image to be recovered adequately. This is equivalent to any other sampling, as defined by Nyquist's Theorem. The image is defined therefore as a frequency domain multiplication of ρ and M, or their convolution in space

$$\lambda'_\phi(x') = \int_{-\infty}^{\infty} \lambda_\phi(x) p(x' - x) dx \tag{21}$$

or more conveniently

$$\lambda'_\phi(x') = \lambda_\phi(x') * p(x') \tag{22}$$

where $p(x')$ is the inverse transform of $|\rho|$.

6.2.4. Applications

Computerised Tomography has been applied to a variety of diagnostic imaging applications. The main area, however, is in determining the extent of cancerous growth since it affords a much higher resolution than projection radiography. The resolution is a result of its ability to discriminate between small changes in absorption coefficient, μ. The spatial resolution of current systems is around 0.4 mm for high contrasts and about 2 mm when the contrast is around 0.5%.

An example of the information which may be obtained from a tomographic scan is shown in Plate 4. The use of helical scanning here has enabled data to be collected through a volume of

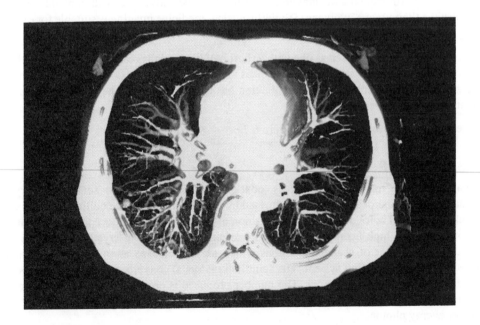

Plate 4 Helical scan of a lung

the patient and then reconstructed to form an image with the illusion of depth. The plate demonstrates the volume affected by cancerous growth hopefully enabling treatment to be adequately planned.

The advantage of higher resolution imaging in these circumstances should be obvious: the greater accuracy of determining the extent of a carcinoma enables its more complete removal by radiotherapy whilst minimising collateral damage to surrounding healthy tissue.

6.3. Gamma Camera

6.3.1. System Outline

A Gamma Camera is used to assist in diagnosing conditions which may be examined by the absorption of labelled materials by organs. Instead of using an external radiation source which is differentially absorbed by materials with varying absorption characteristics (which are largely controlled by the atomic weights of their components), the radiation source is injected into the patient. The source is tagged onto a suitable biochemical reagent which has properties which give it particular affinity to organs whose structure is to be examined. Therefore the emitted photon energy is chosen to be in a band where it is relatively little absorbed, but instead is transmitted. This means that the energy used is normally somewhat higher than is used for projection radiography. It must nevertheless be lower than the energy region which would cause Compton Scattering in tissue. A commercial gamma camera may be expected to be sensitive to energy over the range 50 to 400 keV.

Unlike X ray examinations, nuclear medicine studies are particularly effective at providing differentiation between relatively uniform masses of tissue. Tagged materials are also particularly suitable for studying the movements of fluids in the body, either within blood vessels or to observe their uptake in organs. Similarly, nuclear medicine imaging is suitable for examining moving structures in the body.

The active nuclide is chosen as one with a short half life, and the reagent of a form which will be excreted reasonably readily. These choices minimise the residual dose of radiation which is retained by the patient once the examination has been completed. The choice of indicating materials is discussed below.

The system in outline is shown in Figure 6.7.

6.3.2. Image Receptor

Image reception for Gamma Cameras (or the Anger Camera) is by a scintillating crystal, usually sodium iodide (NaI) which is fronted by a lead collimator. The collimator consists of a matrix of small holes organised in the fashion of a honeycomb to reduce the range of angles of incidence of high energy photons which meet the scintillating crystal. This is done to reduce the number of scattered photons which are detected so as to improve the spatial accuracy of the detecting system. The scintillating crystal converts the high energy photons into a number of visible photons. The number produced is proportional to the energy of the incident photon as it clearly does not release all its energy in a single collision producing a lower energy photon.

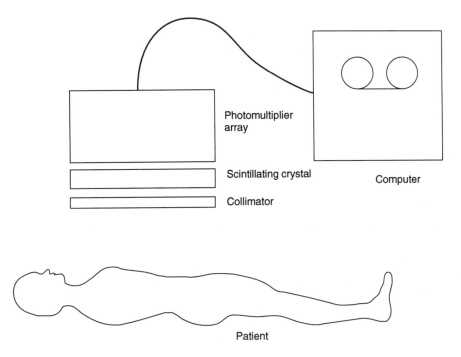

Figure 6.7 Gamma camera schematic

The visible photons are collected by an array of photomultipliers so that they may subsequently be counted. A photomultiplier is a vacuum device whose photo cathode is excited to release electrons when a photon is incident by the photoelectric effect. The released electron is accelerated by a large electric field towards a cathode, but is interrupted on the way by a number of grids. Its collision with each grid causes the grid to release further electrons. The spectral response of the photocathode is matched to the spectral output of the scintillating material. The purpose of this system is to make it possible to count the arrival of single high energy photons which would without the cascade effect of the photomultiplier produce amounts of charge too small to measure.

The output of each photomultiplier could be examined to look at the strength of each pulse received to measure the energy of each incident photon. Only those events which arose from the arrival of photons within the expected range of energies would then be significant for the measurement. The spatial resolution of the instrument depends on the γ photons arriving as single events so that their location may be determined by looking at the relationship between the signals produced by each photomultiplier in the receiving array.

In order to construct an image, an array of photomultipliers is used, arranged packed together over a disc. Their outputs are coupled together in order to ascertain the position of the incident photon on the scintillator. Overall the camera is typically about 400 mm in diameter and 300 mm in length. A modern camera may contain 107 photomultiplier tubes and have a resolution of about 3.5 mm.

The event position is determined electronically by examining the outputs of the set of photomultipliers, using a network of resistors. An estimate of the *x* and *y* co-ordinates where the event took place is obtained by averaging

$$\bar{x} = \frac{\sum_i x_i n_i}{\sum_i n_i} \quad \text{and} \quad \bar{y} = \frac{\sum_j y_j n_j}{\sum_j n_j} \tag{23}$$

In these expressions, the terms x_i and y_j are the co-ordinates of the centres of the photomultipliers, and the n_i and n_j terms their respective counts. A spatial resolution of around 1000 points may be obtained with a resistor network for an array of only 19 photomultipliers. This provides a resolution of around 1 cm which is significantly worse than the resolution of projection radiography. The collimating system effectively controls the resolution of this imaging system. In crude terms this is given by the diameter D of the minimum sized object which may be discriminated as

$$D \approx d(h+L)/h \tag{24}$$

where h is the length of the collimator's holes, d their diameter, and H the distance between the source and collimator. This indicates that using small hole sizes and placing the collimator close to the source both improve resolution. Hole sizes for 400 mm diameter Gamma Cameras are in the range 2 –3 mm diameter.

Other forms of Gamma Camera are in use. They use either circular array of detectors or pairs of receptors which enable the collection of depth information from nuclear medicine examinations.

6.3.3. Isotope Choice

As was mentioned previously, the isotope used in nuclear medicine has to produce photons of a suitable energy. These should normally be of somewhat greater than 100 keV so that they are not absorbed to the same degree as would be lower energy photons, and do not suffer from Compton scattering to the extent that would photons of significantly higher energies. Further, it is obviously desirable to choose an isotope which has a half life of only a few hours so that the exposure the patient receives does not significantly exceed the period of examination. Equally the isotope when it decays should not be significantly active in its new form before it has been thoroughly voided, and it should not cause significant toxic effects. Owing to the desire for short isotope lifetimes, any materials must be able to be made in a laboratory close to where the patient will be examined. The isotopes used are therefore the immediate offspring of longer lived nuclides which may be chemically separated.

The commonest nuclide used in nuclear medicine is $^{99}\text{Tc}^{\text{m}}$ (technetium) which is able to bind to a range of biological reagents. Its decay reaction is

$$^{99}\text{Mo} \xrightarrow{\beta^-} {}^{99}\text{Tc}^{\text{m}} \rightarrow {}^{99}\text{Tc} + \gamma$$

This means that the parent, Mo (molybdenum) decays by the loss of a nuclear electron (β decay) to produce a metastable form of Tc, which loses its recoil energy as a γ. The first decay takes place with a half life of 66 hours, and the second with one of six hours. The Mo source itself is produced either as a by-product of the fission of uranium or by thermal neutron capture with a lighter isotope of Mo.

The patient dose as a result of a nuclear medicine investigation is very much dependent on the nuclide used. Its half life clearly determines how long it may remain active, but the result of the dose received also depends on how much it is localised on one organ as a result of the tagged material being taken up by that organ, and then the rate at which that material is metabolised and excreted. Typical doses are in the range 0.1 to 2 mGy; thyroid doses of up to 500 mGy are likely when radioactive iodine is used.

6.4. Nuclear Magnetic Resonance Imaging

NMR scanning affords the possibility of scanning volumes of the body in order to map the density of hydrogen nuclei. A map of protons in the more loosely bound water molecules in tissue can give a good view of the structure of the tissue. The images produced by NMR are of very good contrast and high resolution. The apparatus permits the generation of sectional images in the same manner as CT, but without the attendant dangers of ionising radiation. The physical basis of the magnetic resonance phenomenon was outlined in Chapter 3 above.

NMR spectroscopy is carried out using a very strong magnetic field. This in itself is not known to be particularly hazardous (see the exceptions below), although the field strength is sufficient to exert very strong forces on any ferro magnetic materials, such as surgical implements, metallic body implants, watches or the like. These may be dangerously accelerated in the room used for examination, and must be carefully excluded at all times.

There is also a danger from the field in that it could generate an induced emf in the blood stream as a result of the passage of ionic material in the blood which is in the right sort of range to induce depolarisation voltages in the heart. The danger of the static field is, however, less than that of sudden removal of the field. There are similar hazards as a result of its sudden removal during an emergency shutdown causing a massive rate of change of flux. The apparatus must be designed to prevent its shutdown from occurring too quickly even on power outage.

The magnetic field is of sufficient strength for it to require to employ superconducting magnets in circumstances where field strengths of greater than 0.3 T are required (compare with the strength of the earth's field at about 50 μT). At the higher fields then the superconducting magnet requires to be maintained at liquid helium temperatures (4 K). This causes significant problems in the design of the magnet so that it may produce a sufficiently accurate field over the volume required for medical measurements whilst at the same time retaining its thermal insulation to ensure that the helium remains liquid.

6.4.1. Image Creation

The creation of an image from information derived from the NMR effect depends upon determining the spatial distribution of resonating particles. As we have seen, the resonant frequency is dependent on the strength of the magnetic field **B** which is used to line up the precession of momentum. If we vary the strength of that field over the region of interest, then particles in the region will have different resonant frequencies. A stimulating impulse which transmits magnetic energy to them of the resonant frequency causes members of the population to resonate. The decay of their resonance may then be measured by the use of a magnetic sensor (a coil).

Recall from Chapter 3 that resonating nuclei decay from their coherent resonance condition in a time. This is due to their resonant energy being either returned to the lattice or exchanged with other nuclei. These two processes were characterised by the relaxation times T_1 and T_2 respectively.

If we wish to excite a group of nuclei experiencing a magnetic field **B** into a resonance condition, then they can be made to precess about the field if excited by a rotating field at their resonant frequency. The fastest manner to sweep through a sample of material in order to locate the concentration of resonating nuclei is to provide a short burst of energy at their resonant frequency. The Fourier Transform of a square impulse was shown in Chapter 5 to be the sinc function. Owing to the symmetry of the transform and its inverse, this is the shape of the envelope function required to modulate the carrier frequency if we are to obtain a simple pulse of energy at the resonant frequency without sidebands. Unfortunately the sinc function itself is of infinite bandwidth, but it may be adequately approximated within the bandwidth of our measurement system by truncation. The distorting effect of the bandwidth limitation may be ameliorated by the use of a damping function.

We now require to select components of our sample. We do this by imposing a gradient in the magnetic field in the direction of the main magnetic field. An outline of the apparatus and conventional directions we use is shown in Figure 6.9. This gradient means that each plane normal to the gradient has a different resonant frequency. The excitation pulse contains frequencies mainly in a small band: these then excite the set of planes with resonant frequencies in that range. The gradient field and its effect on the frequency of resonance are shown in Figure 6.8.

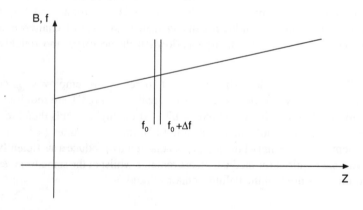

Figure 6.8 Gradient fields used to select region for excitation.

The next step is to attempt to differentiate between resonating nuclei in the slice normal to the z-axis. This may be done by removing the z-gradient and instead imposing a gradient in another direction. For instance the application of a gradient along the x-axis enables the selection of nuclei which received energy from the previous excitation pulse. Their frequencies of resonance are selected by the resulting field including the new gradient. The spectrum of frequencies of resonance detected then represents the individual projections. This is analogous to the procedure we discussed earlier (in section 6.2.3) when we used a Fourier method to reconstruct a tomographic image.

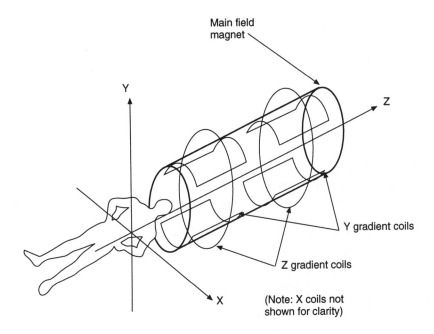

Figure 6.9 Schematic representation of NMR scanner

6.4.2. Application of MRI

The use of magnetic resonance scanners in medical diagnosis is limited substantially by their cost. The systems require intrinsically expensive apparatus and high performance computers to create images. Owing to their size and weight they may require to be sited in special rooms which can tolerate the high mechanical loads imposed. They do, however, provide images of a previously unobtainable quality as they rely on a physical process which is not accessible to other imaging modalities. Plate 5 is a picture of a currently available commercial scanning system.

The scanners inherently obtain three dimensional image data. This may be presented as a longitudinal section through the body, as in plate 6 which shows a sectional view through the brain. A view with simulated perspective may also be obtained, as in Plate 7 which should be contrasted with the image obtained from the helical computerised tomography scan shown in Plate 4. Finally a view may be obtained by processing to provide a surface map of the brain to assist in surgical planning as shown in Plate 8.

Image processing coupled with MRI scanners may be used to recognise anatomical features, such as cardiac structures, and then track their movement. This enables both cardiac stroke volume to be measured and other anatomical details such as the cardiac wall thickness to be monitored.

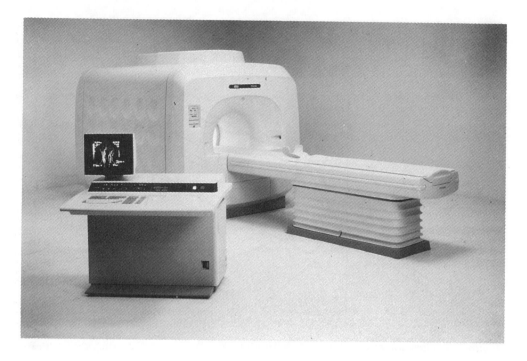

Plate 5 Commercial scanning system

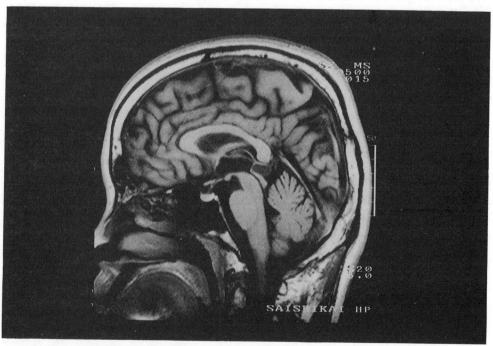

Plate 6 Longitudinal section through the brain

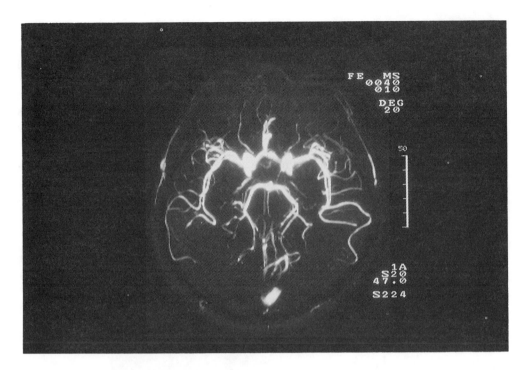

Plate 7 Simulated perspective

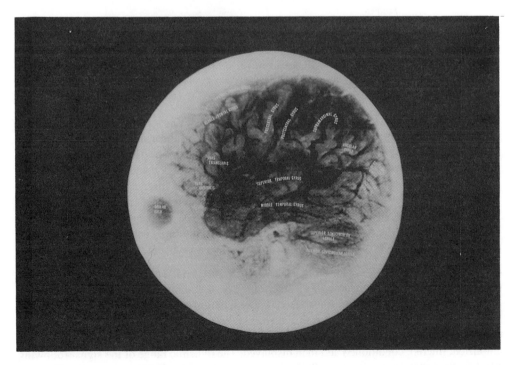

Plate 8 Surface map of the brain

6.5. Ultrasound Imaging

Ultrasound provides in its simplest form the means to obtain a limited amount of structural information without exotic technology and with a good degree of patient and operator safety. The following sections describe firstly the simplest application of the technology which underlies most of the systems employed. We go on to present a description of the systems which employ more sophisticated technology to enable the production of images which may be more readily interpreted.

Whilst there are some doubts about the routine use of ultrasonic imaging in pregnancy, the present evidence of hazard is disputed. The technique would appear to be inherently reasonably safe: we explore the safety of ultrasound further in Chapter 8.

6.5.1. A-SCAN

The A-scan is the simplest form of ultrasonic scanner. It is a pulse echo or Time of Flight (TOF) imaging system, in that the time for a signal emitted from a transducer to return is related to the distance it has travelled.

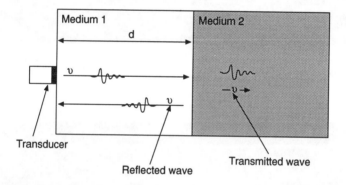

Figure 6.10 Transducer object reflection

In Figure 6.10 an ultrasonic pulse is transmitted by the transducer into medium 1. It travels until reaching medium 2 whereupon part of the wave is reflected and the remainder transmitted. The reflected wave travels back through the medium and is detected by the transducer. Therefore a return echo from the medium arrives time t after the pulse is generated corresponding to the pulse having travelled distance $2d$ (there and back). In an imaging situation d is not known, but can be calculated if the velocity of the propagating wave is known. Therefore assuming the velocity of the medium is v then a returning echo t seconds after pulse emission corresponds to an interface located $vt/2$ metres away from the transducer.

A schematic of a basic A-scan system is shown in Figure 6.11. Pulses are produced at a rate (the Pulse Repetition Frequency, PRF) determined by the impulse generator. The pulses are then amplified and excite the transducer. The excitation signal is usually an impulse of between 200 V and 300 V. One transducer may be used for both signal generation and detection or alternatively two separate transducers may be used. The return signals are amplified, filtered, conditioned and then displayed. The dynamic range of the signal following amplification may be as high as 100 dB. The return echoes are displayed on an oscilloscope

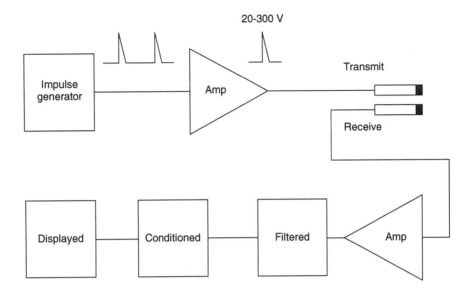

Figure 6.11 Schematic of A-scan system

whose sweeps are triggered by the pulse generator at a rate governed by the pulse repetition frequency.

6.5.1.1. The pulse repetition frequency

The PRF determines the rate at which pulses are emitted from the transducer. For a maximum display intensity this rate should be as high as possible.

In Figure 6.12 a pulse is transmitted at time zero, causing reflections which will be received from the interfaces situated near to the transducer almost immediately and further reflections from distant interfaces up to time t seconds later. The echo at t_2 is the echo from the farthest interface. Therefore if a second pulse were emitted by the transducer before t_2 then the transducer would detect echoes associated with the first pulse whilst also receiving echoes from the second pulse. These late echoes from the first pulse would be highly attenuated owing to the distances travelled but could be mistaken for weakly reflecting interfaces near to the transducer. Therefore for unambiguous detection a second pulse must not be emitted until all possible echoes have been received from the furthest possible interface. This sets a limit to the maximum pulse repetition frequency

$$PFR_{max} = \frac{v}{2d}$$

where v is the wave velocity in the medium and d is the furthest reflecting interface.

6.5.1.2. Swept gain control

When an echo returns from a distant interface it is considerably smaller than an echo returning from an interface closer to the transducer. This is due to the attenuation of the wave by the tissue through which the pulse travelled. To compensate for this tissue attenuation the gain of

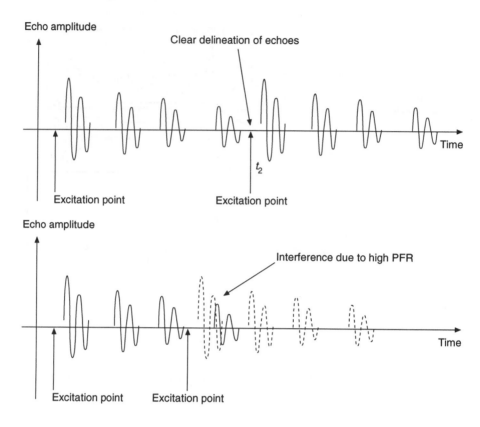

Figure 6.12 Pulse interference due to high PRF

the system is varied. Pulses entering the transducer from interfaces close to the transducer are attenuated with respect to pulses entering from a greater depth. Figure 6.13 demonstrates the principle. This method of accounting for the tissue attenuation is referred to as a Swept Gain function. Swept Gain also helps reduce the likelihood of echoes from an earlier pulse whose return time was greater than t_2 being observed, as the system gain is reduced to below the level at which they may be detectable. The swept gain function has a dead time in which no echoes are displayed. This obscures echoes which originate from interfaces close to the probe or from the previous pulse. Generally, reflection from close to the probe (due to skin and subcutaneous fat layers) is of no interest. The swept gain section decreases the dynamic range of the detected signal between approximately 100 dB and 50 dB.

6.5.1.3. Display

The cathode ray tube used for A-scan display has its time base calibrated in tissue depth. Basic A-scan systems display the return echoes with no signal processing or detection. However, the majority of scanners also provide half and full wave rectified and smoothed displays. The three methods of display are depicted in Figure 6.14. The full wave rectified display simplifies interpretation of the data while observation in all three modes may be required to separate multiple echo signals.

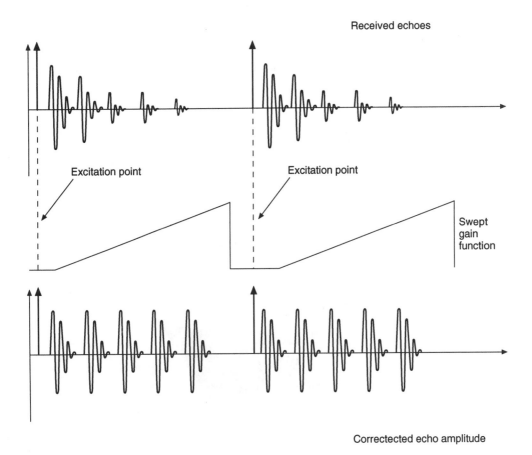

Figure 6.13 Swept gain control

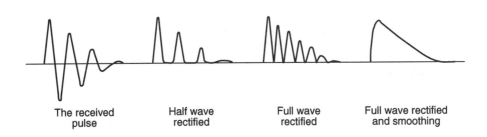

Figure 6.14 Half and full wave rectified echoes

6.5.1.4. Imaging problems

Ultrasound imaging systems rely on the assumption that the time of flight is directly related to the depth of the reflecting interface in a straight line from the probe and that the tissue propagates ultrasound uniformly. However, the ultrasonic velocity is not constant in the body as the differing tissues have different material properties. In addition ultrasound may return to the transducer from an interface off the axis of the beam following multiple reflections.

1. If an ultrasonic pulse (see Figure 6.15a) is reflected from an interface away from the transducer, the pulse suffers a second reflection back towards the transducer. Calculation of the depth of the interface from the time of flight will indicate an interface at a greater depth than the first interface.

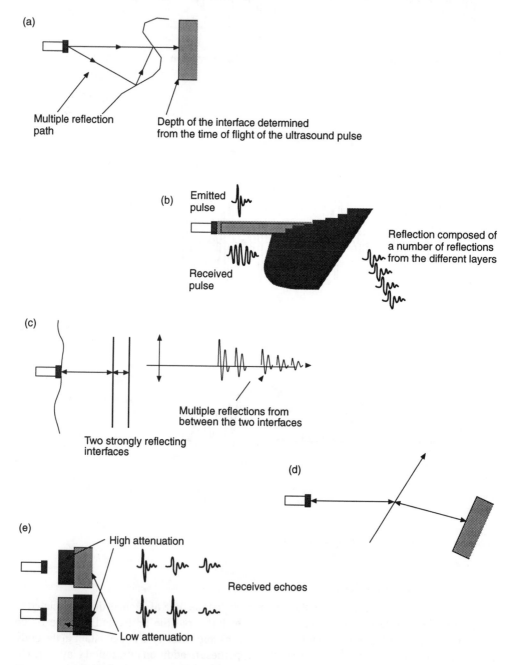

Figure 6.15 Problems in A-scan imaging

2. If an ultrasound pulse is reflected from an interface which is stepped at an angle to the beam, due to the finite width of the ultrasonic beam an elongated pulse will be detected. This will make detection of the depth of the interface uncertain (see Figure 6.15b).

3. An ultrasonic probe could be positioned on the patient's tissue above a pair of strongly reflecting interfaces one behind the other. A pulse travels to the first interface and a proportion is reflected back to the transducer. The transmitted wave continues and is reflected by the second interface. The echo from the second interface partially reflects off the first interface and back to the second interface. This process may continue with the pulse bouncing back and forth between the two interfaces. At each reflection a proportion of the ultrasound is transmitted back to the transducer and is detected. The A-scan will therefore detect a series of echoes from the two interfaces (see Figure 6.15c).

4. A pulse travelling through a refracting medium is bent. If it then impinges upon a strongly reflecting interface the returning echo is once again refracted and returns to the transducer. The depth of the interface calculated from the time of flight will indicate an interface at a depth equal to the trip distance but along the axis of the ultrasound beam (see Figure 6.15d).

5. If as depicted in Figure 6.16 there is a region of high attenuation preceding an interface, then echoes from beyond that region will appear weaker than normal. Alternatively if there is a region of low attenuation preceding an interface then echoes from beyond that region may appear stronger than normal. In both these situations the body has non uniform attenuation which alters the relative size of the echoes (see Figure 6.15e).

6.5.1.5. Axial resolution

The axial resolution of an ultrasound scanner defines its ability to differentiate between two reflectors on the same axis but separated by a displacement d. Clearly this is largely determined by the bandwidth of the transducer, its excitation signal and the detection circuitry.

Figure 6.16 shows the theoretical response of two transducers in both time and frequency domains. Transducer (a) is a high Q lightly damped transducer, its time domain response shows considerable ringing (continuing oscillations). The response of transducer (b) is wide bandwidth and highly damped with a low Q. Figure 6.16 shows the signal received when the two transducers are used to detect two reflectors. The highly damped transducer will be able to differentiate between the two reflectors at a closer separation than the lightly damped transducer. This example demonstrates that for pulse echo imaging the axial resolution is dependent on the time domain response of the transducers used. The excitation pulse must also have a wide bandwidth to elicit the optimal response from the transducer. Typically the excitation pulse is in the order of 100 ns wide.

6.5.1.6. Interpretation

The A-scan produces images of all the reflecting structures within the ultrasound beam. If a number of interfaces are close together it may be difficult to separate them. The human body is a very complicated structure and therefore considerable skill and experience is required to interpret the scan. In instances where the expected composition of the body is not known the scan may be almost impossible to interpret.

6.5.1.7. Modern uses

The A-scan has been largely replaced by the B-scan. However, there is a number of areas where structures are simple enough to be interpreted.

1. *The eye*

   The A-scan still finds application in ophthalmology as the eye is a simple structure and therefore the A-scan can be interpreted. An A-scan probe is placed on the end of a water-filled tube in contact with the eye (see Figure 6.17). The scan is then performed to detect foreign bodies within the optical cavity.

2. *The mid line of the brain*

   Following a trauma or perhaps with no prior symptoms, a patient may develop bleeding within the skull. The brain is divided in two along the mid sagittal plane (see Figure 6.17),

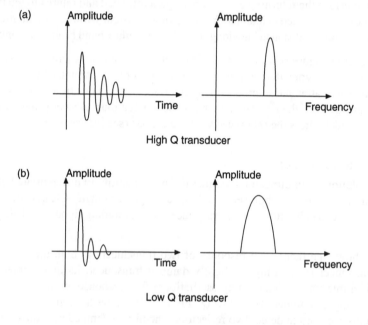

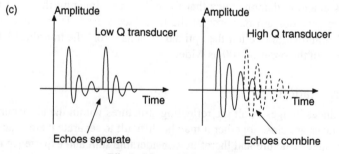

Figure 6.16 Transducer frequency and time response

A–Scan of The Mid Line of The Brain

A–Scan of The Eye

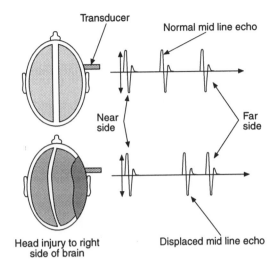

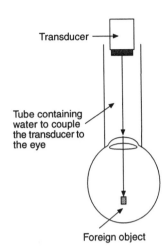

Figure 6.17 Eye and mid line scan

which is referred to as the mid line of the brain. In a case of internal bleeding the build up of pressure on one side may cause the mid line to become displaced. This displacement can be detected using an A-scan. In this situation again the object under test has a simple internal structure as the mass of the brain will appear to be homogeneous.

Away from the medical field the A-scan is used extensively in Non Destructive Testing (NDT), for detecting cracks in uniform materials such as steel and in quality assurance in detecting the dimensions of materials. In both of these applications the A-scan device is ideal, as the structures are simple and therefore the echo display is easy to interpret.

6.5.2. B-scan

B-scan imaging systems essentially consist of an A-scan device which is physically swept across the patient's skin. At each position an A-scan is performed, the amplitude of the reflection from the various interfaces within the patient is then used to modulate the brightness of a line on an *x-y* display. Each separate line is formed by a different A-scan. In this way a picture of a section through the patient is developed allowing the shape of the internal organs to be recognised.

In Figure 6.18 the ultrasound probe is used to transmit a pulse. The resulting echoes from the interfaces shown are full wave rectified and used to determine the brightness of the display along one line. The ultrasound probe is then moved to scan along the adjacent line and the echoes are again used to determine the brightness.

A schematic of a basic B-scanning system is shown in Figure 6.19.

The system essentially consists of an A-scan system with the addition of :

• position information from the probe.

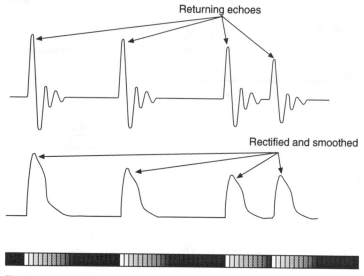

Returning echoes

Rectified and smoothed

The brightness of the display coordinates along a line are determined by the echo amplitude along that line

Figure 6.18 B-scan image formation

- a range compression section.

- a display section which combines the position information and the echo signal to form a line on the display.

The position of the probe is determined and fed to the display section where the echo from each position is used to generate the x-y display. The detected echo signal following the swept gain section has a dynamic range of approximately 50 dB (the ratio of the weakest to the strongest echo). The CRT has a dynamic range of approximately 20 dB, hence the echo signal is compressed to allow display. This is achieved by using a non linear amplifier, whose gain is decreased as the signal amplitude increases. This may be implemented with a logarithmic amplifier.

6.5.2.1. Movement Artefact

The first B-scan systems developed produced a picture over approximately five minutes; during this time the internal organs of the patient and the patient himself may not have remained still. Therefore the images obtained were considerably distorted. However, the scanning rate of B-scan systems has improved to the extent that the picture may be updated 30 times a second allowing clear flicker free images and imaging of moving structures.

The rate at which a B-scan can be performed is determined by the depth to which the scan is required. In the A-scan the pulse repetition frequency was limited by the depth of the scan as a second pulse could not be generated until all the echoes from the first scan had returned. In the B-scan the scanner has to wait for the echo to return from the deepest organ of interest before scanning the adjacent line. The entire image must be updated 30 times a second for flicker free imaging. Therefore, the pulse repetition frequency is fixed for a given depth and

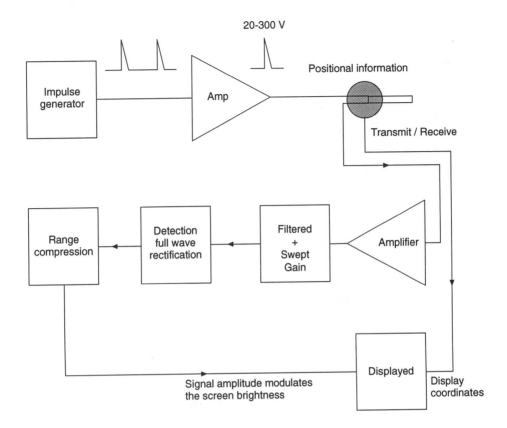

20-300 V

Positional information

Impulse generator

Amp

Transmit / Receive

Range compression

Detection full wave rectification

Filtered + Swept Gain

Amplifier

Displayed

Display coordinates

Signal amplitude modulates the screen brightness

Figure 6.19 Schematic of B-scan system

the time for capture of the whole image is also fixed. These two fixed relationships determine the number of lines which make up the B-scan image. Hence

$$\text{time for one line scan } = \frac{d}{v}$$

where d is the depth of the deepest organ of interest and v is the velocity of the wave in the tissue.

Therefore the number of lines which make up one scan is given by $v/R \times d$ where R is the screen refresh rate.

6.5.2.2. Sector Scan, Transitional Scan

There are two forms of scan: the sector scan and the transitional scan (see Figure 6.20).

In a sector scan the transducer is rotated or rocked about an axis to obtain a sequence of A-scan lines at differing angles. The sector scan is probably the most common type of scan. The rotating motion is easily achieved and a large internal area of patient can be examined from a small surface area in a signal position. Sector scanning also avoids possible problems with scanning through a non uniform area of the patient's body as the picture is effectively obtained through a single point.

In the transitional scan the transducer is moved in a line to produce a rectangular shaped picture. The picture is easier to interpret, but access to a larger area of patient is required and this makes transitional scanning relatively clumsy.

If the ultrasound beam meets an interface at an angle the ultrasound may be reflected at an angle away from the transducer. In this instance no signal will return to the scanner and consequently the interface will not be imaged. This situation can occur with both sector and transitional scanning. To overcome this systems have been developed that perform both translation and sector scans or a series of sector scans at different translation positions. In this way the interface is imaged from two different angles reducing the possibility of reflections not reaching the probe. With scanners of this kind the image is developed as the summation of the echoes from different angles relating to the same patient position.

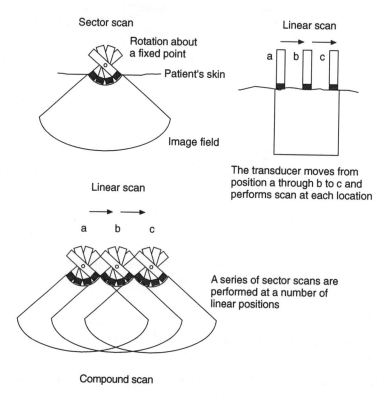

Figure 6.20 Sector / transitional scan

6.5.2.3. Transducers for B-mode imaging

1. Fixed focus transducers

The lateral resolution of a transducer can be improved by focusing the transducer. This can be achieved by either using a curved transducer substrate or using an acoustic lens. The two situations are depicted in Figure 6.21. Transducers fabricated in this way do not have a focal point as such, but are focused over a region referred to as the focal zone. However, the lateral resolution beyond the focal zone deteriorates rapidly.

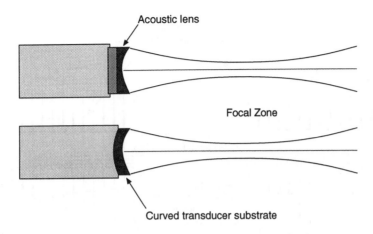

Figure 6.21 Transducer focussing

2. Linear Array Transducers

Early transducers for B-mode scanning were moved manually and the position information fed to the display section. Subsequently transducers were moved by motors. However, transducer development has advanced to make physical movement of the transducer unnecessary.

A linear array transducer is shown in Figure 6.22, which comprises approximately 150 different transducer elements. A group of approximately 10 elements are excited simultaneously and function as a single transducer. These elements are then used as a group to detect the reflected echoes. Following this an adjacent set of transducer elements are excited to perform the scan for the following line. In this way a translation scan can be performed. If the linear array transducer is formed on a curved substrate then by following the same procedure a sector scan can be performed. See Figure 6.22.

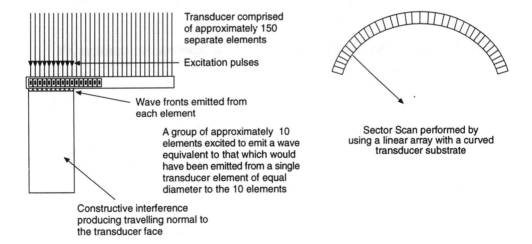

Figure 6.22 linear array transducer

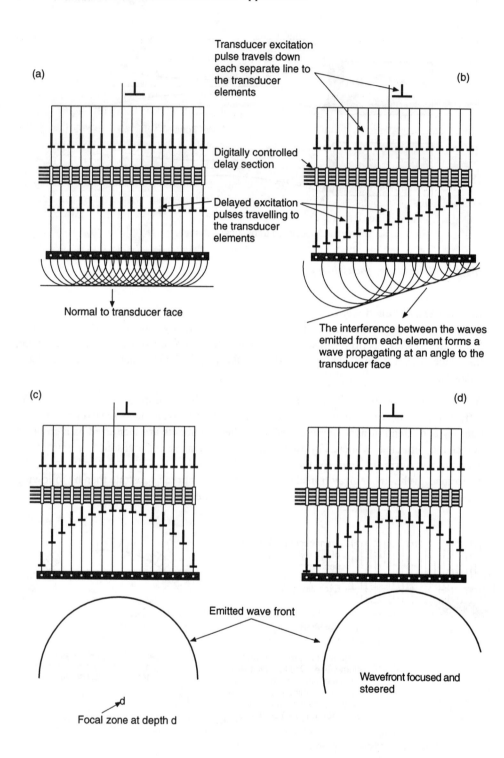

Figure 6.23 Phased array transducer

The substrate upon which a linear array transducer is fabricated is curved in the plane orthogonal to that of the scan to allow a degree of focusing.

3. **Phased array transducers**

    A phased array transducer consists of a series of transducer elements as in the linear array each of which can be fired separately. If the excitation signal were fed to all elements at the same time the transducer would behave as a single large transducer. However, the excitation signal is delayed in a carefully chosen manner to achieve transducer focusing and beam steering.

    Figure 6.23a shows the excitation pulses reaching transducer elements and the subsequent emitted ultrasonic pulse waveform. In Figure 6.23b the pulse is steered to the right by using a linearly incremented delay between each transducer element. Similarly in Figure 6.23c the transducer element delay is configured to produce a beam which is focused to a depth *d*. Figure 6.23d shows a combination of b and c producing a focused beam directed at an angle.

    The phased array can also be focused and steered to be maximally sensitive to echoes returning from a particular angle: this situation is demonstrated in Figure 6.24. The delay of the received echo from each element is adjusted so that a wave entering the transducer from the required angle will experience constructive interference while a wave entering the transducer at normal incidence will experience destructive interference

    Using a phased array the ultrasonic beam can be steered through a range of angles to produce a sector scan and can be focused at any range of interest.

Electronic focusing of an ultrasonic transducer is effective in one plane only and so the transducer is formed from a curved crystal to produce fixed depth focusing in the other plane.

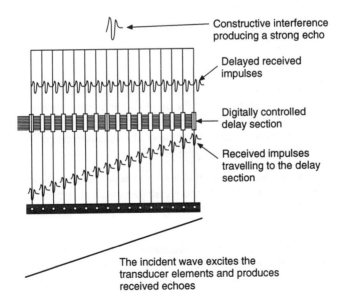

Figure 6.24 Phased array receiving

6.6. Doppler Ultrasound

6.6.1. Introduction

In section 3.8 we saw that when ultrasound was scattered by a moving object or interface its frequency was altered. The change in the frequency was directly proportional to the velocity of the scatterer, acting along the axis of the ultrasound beam. The equation defining the frequency shift was:

$$f_d = \frac{\pm 2 f_s V \cos\theta}{c}$$

This principle is made use of in the Doppler blood flow meter. When an ultrasound beam is incident on a blood vessel a proportion is scattered back along the incident path towards the emitting transducer. A Doppler flow meter detects the frequency variations in back scattered ultrasound and produces a proportional output. Doppler flow meters are used to assess the integrity of the circulation system. There is a variety of other medical devices which employ the Doppler principle, namely the foetal heart rate monitor, foetal motion and breathing detector and blood pressure measurement.

6.6.2. The Origin of the Scatter Signal

Blood has three main constituents, platelets, leukocytes and erythrocytes which are suspended within plasma. Initially it was thought that incident ultrasound was scattered by the individual cells within the plasma. However, models treating cells as individual scatterers failed to predict the scatter observed. More recent theories consider the scatter to originate from variations in the compressibility and density of blood due to pressure fluctuations during the cardiac cycle.

6.6.3. Continuous Wave Doppler Instrument

The first and probably the most commonly used Doppler device is the continuous wave flow meter shown below in Figure 6.25.

The oscillator produces a sine wave of the required frequency which is amplified and fed to the transducer. The transducer is driven by the amplifier at approximately 10 volts, in sharp contrast to the hundreds of volts used to excite transducers in imaging applications. In the continuous wave flow meters the signal is continuously transmitted and so high peak levels are not required to produce the required signal power.

As the ultrasound signal is continuously transmitted a separate transducer is used to detect scatter. The received signal is then fed to a high frequency amplifier and demodulated. The demodulated signal is then either detected aurally by the operator or its frequency variations are detected and displayed.

The continuous wave flow meter detects frequency shifts from any moving scatterer or interface in the ultrasonic beam. Therefore, frequency shifts are detected from the relative movement of the probe and the patient and from moving interfaces within the body such as organ or blood vessel walls.

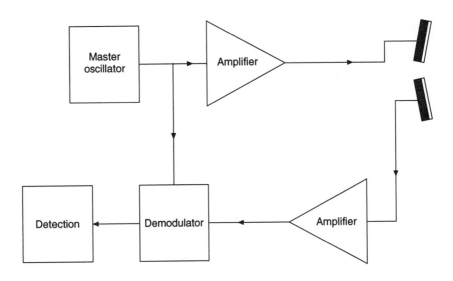

Continuous Wave Doppler System

Figure 6.25 (CW flow meter)

6.6.3.1. Transducer operating frequency

Doppler transducers are designed to resonate. The piezo electric element within the trans-ducer is air backed allowing an un-damped high Q response. The Doppler probe consists of both transmitting and receiving transducers, which may be fabricated on the same piezo electric element.

The Doppler shift is directly proportional to the ultrasound carrier frequency. Therefore, to obtain a maximised frequency shift, it would be advantageous to use a high transmission frequency. Unfortunately the attenuation of ultrasound increases with increasing frequency, although the amount of scatter from the blood increases with frequency to the fourth power. Hence, when determining the ultrasound frequency for a particular application these three factors must be considered. Figure 6.26 indicates the application for a range of frequencies of ultrasound and operating depth.

Clinical Application	Depth of focused field	Ultrasound Frequency
Obstetric	70–100 mm	2 MHz
Cardiovascular	20–30 mm	4 MHz
Peripheral Vascular	8–15 mm	8 MHz
Opthalmic / Peripheral	7–10 mm	10 MHz

Figure 6.26 Beam overlap

To obtain a degree of range discrimination the acoustic fields of the transmitting and receiving transducers overlap at the intended scanning depth. The transducers are focused for this depth. This reduces the interference from moving interfaces outside the intended region.

6.6.3.2. Demodulation Techniques

The back scattered Doppler signal is centred at the frequency of the transmitted ultrasound (sometimes referred to as the carrier frequency). It is possible to detect the Doppler frequency variations by processing signals directly at this frequency. The system required is simplified if the signal is modulated to a lower frequency. All commercially available Doppler flow meters mix the Doppler signal with another high frequency to obtain a lower frequency shift signal.

The Doppler signal at the carrier frequency contains information about the velocity of the scatterers and their direction. The Doppler shift is either positive or negative equating to flow towards or away from the ultrasound probe. In some Doppler imaging situations it is advantageous to determine the direction of flow. Doppler demodulation techniques can be considered as either preserving or destroying this directional information.

6.6.4. Modulation

Multiplication of a signal by a sinusoid shifts the original message by the frequency of the sinusoid: the process is known as amplitude modulation. Amplitude modulation can be explained by examining the resulting spectra.

If we define a signal $g(t)$ such that

$$g(t) \leftrightarrow G(\omega)$$

The multiplication of $g(t)$ by $e^{i\omega_0 t}$ is equivalent to a frequency shift of ω_0. Hence:

$$g(t)e^{i\omega_0 t} \leftrightarrow G(\omega - \omega_0)$$

The modulating signal $\cos \omega_0 t$ in exponential notation is:

$$\cos \omega_0 t = \frac{1}{2}\left(e^{i\omega_0 t} + e^{-i\omega_0 t}\right)$$

so multiplying, we obtain

$$g(t)\cos \omega_0 t = \frac{1}{2}\left(g(t)e^{i\omega_0 t} + g(t)e^{-i\omega_0 t}\right)$$

then

$$g(t)\cos \omega_0 t \leftrightarrow \frac{1}{2}\left(G(\omega + \omega_0) + G(\omega - \omega_0)\right)$$

Hence multiplication of $g(t)$ by $\cos \omega_0$ produces a scaled copy of the message signal centred at $\omega \pm \omega_0$. For a fuller explanation, see, for instance, Lathi (1983).

6.6.5. Non Directional Demodulators

The simplest form of signal demodulation is direct multiplication with the oscillator signal. This form of modulation shifts the signal by the carrier signal producing a copy of the signal centred about the axis. This produces a low frequency copy of the signal. However, this method of demodulation loses the directional information as the demodulated signal is composed of both forward and reverse flow information.

6.6.6. Directional Demodulators

The three commonly used demodulation methods which preserve directional information are quadrature, side band and heterodyne.

1. In quadrature demodulation the back scattered signal is fed to two demodulators. One demodulator is also fed with the carrier signal whilst the other is fed with the carrier signal phase shifted by 90°. This process produces two demodulated channels referred to as the in phase and quadrature channels respectively. Each channel contains forward and reverse flow information.

2. In side band demodulation the back scattered signal is fed to two sections each of which contains a filter and a demodulator. In one section the signal is filtered by a high pass filter centred at the carrier frequency and then demodulated by the carrier. In the other section the signal is fed to a low pass filter centred at the carrier frequency and then demodulated. This produces two channels one of which contains the forward flow information with the other containing the reverse flow information.

3. The heterodyne demodulator multiplies the back scattered signal by a sine wave of lower frequency than the carrier. This produces a copy of the Doppler signal situated at the difference frequency between the carrier and the demodulation frequency. For instance if the demodulator signal is 20 kHz lower than the carrier then the Doppler signal will be centred at 20 kHz. The signal is at a convenient low frequency for detection.

6.6.7. Detection Techniques

Following demodulation the Doppler waveform is detected to determine the frequency information. The simplest method of achieving uses a zero crossing detector.

The zero crossing detector relates the number of times the Doppler signal crosses the zero line to the instantaneous frequency. In a Doppler system there is significant noise. The zero crossing detector therefore is set to detect crossings of a threshold level above zero, so that no crossings are detected when there is no received Doppler signal.

The Doppler signal originates from a number of scatterers moving with a range of velocities; hence the zero crossing detector produces an output which is effectively the RMS value of the Doppler frequency at any instant. The performance of zero crossing detectors is poor in the high noise environments encountered by Doppler flow meters.

Developments in computer hardware have made real time computation of the frequency spectra of the Doppler signal possible. The Doppler back scattered signal is transformed to the frequency domain using a Fast Fourier Transform (section 5.2.2). This method of detection allows the contribution from all scatter to be displayed rather than the average or RMS value of the waveform. The computation is performed by computer or specialised Digital Signal Processing hardware. The three co-ordinates which need to be simultaneously displayed are time, frequency and spectral amplitude. The display x axis correlates with time and the y axis with frequency: the spectral amplitude is displayed as the brightness or the colour of the co-ordinate.

6.6.8. Pulsed Wave Instrument

Using a Continuous Wave or Doppler flow meter the blood flow from a vessel located behind another vessel moving interface cannot easily be studied. Despite the discrimination provided by the transducer configuration the Doppler signal from unintended scatter cannot be separated from the desired target. This severely limits the range of application of Doppler flow meters. To overcome this the pulsed Doppler flow meter has been developed.

Pulsed Doppler flow meters use both the ideas of ultrasonic pulse echo imaging and the Doppler flow meter. A signal is transmitted in a short burst. Signals originating from the required tissue depth are selected by their time of flight. The return signal is then demodulated and detected.

The pulsed Doppler flow meter provides information about the blood velocity from a defined depth. This is achieved by transmitting a pulsed signal rather than a continuous wave. A schematic of a pulsed Doppler flow meter is shown below in Figure 6.27. The meter basically combines the A-scan and the continuous wave flow meter.

Gated sine waves are generated and following amplification are fed to the transmitting transducer. The sine wave pulse is emitted and travels through the tissue. As the pulse travels ultrasound is reflected back to the receiving transducer. The received ultrasound is frequency shifted if the interface or scatterer moves relative to the probe.

Pulsed Wave Doppler System

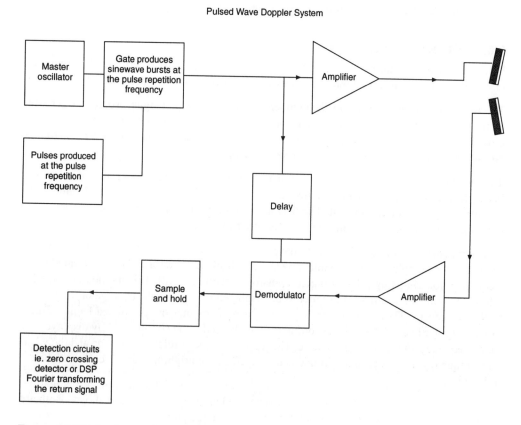

Figure 6.27 Pulsed wave flow meter

The signal received time t after the emission of the pulse originates from a tissue at a depth defined in section 6.5.1. Time t after the emission of the pulse the gated sine wave is fed to the demodulator. Therefore the signal originating from the required depth of tissue is demodulated. The output of the demodulator is zero except when the demodulator pulse is received. Therefore the demodulator output represents the Doppler signal from the required tissue depth after t seconds. The output of the demodulator is fed to a sample and hold circuit, which maintains the level until the next pulse is transmitted to the demodulator. In this way the output of the sample and hold amplifier is updated at the pulse repetition frequency.

The frequency variations of the demodulated wave are detected using the methods described for continuous wave instruments.

6.6.9. The RangeVelocity Ambiguity Function

The pulsed Doppler flow meter effectively samples the velocity of blood at the pulse repetition frequency. The Doppler waveform obtained has a frequency content determined by the velocity of the blood. The Nyquist Sampling Theorem states that the sampling rate must be a minimum of twice the maximum frequency component of the sampled signal. Therefore, the pulse repetition frequency (PRF) must be twice the maximum frequency of the Doppler waveform (i.e. the sample and hold amplifier output).

The maximum frequency that can be unambiguously determined is equal to half the PRF therefore

$$\mathrm{PRF} = \frac{4 f_s v \cos\theta}{c}$$

Recalling that the PRF is given by:

$$\mathrm{PRF} = \frac{c}{2d}$$

it follows that

$$\frac{c}{2d} = \frac{4 f_s v \cos\theta}{c}$$

and hence

$$\frac{c^2}{8 d f_s} = v \cos\theta$$

6.6.10. The Duplex Doppler Scanner

The Duplex Doppler scanner combines a B-scan system with a pulsed Doppler flow meter. Clinically the operator is able to view a section of a patient using the B-scan system. If a blood vessel is identified for investigation, the operator can set the co-ordinates from the B-scan image to perform a Doppler A-scan. A Duplex Scanning System is shown in Figure 6.28. The pulsed Doppler investigation may be performed with the same transducer as the B-scan. However, this necessitates that the B-scan is interrupted for the duration of this investigation. Therefore modern systems have separate B-scan and Doppler transducers as shown. Both are phased arrays which can be directed and focused within the patient.

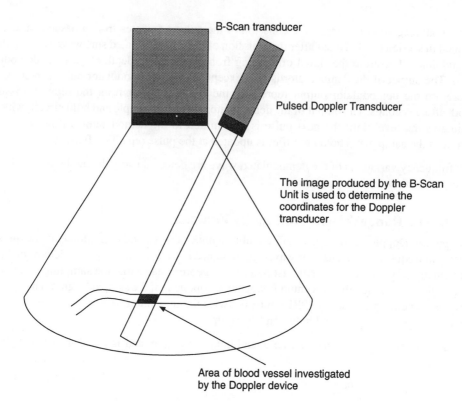

B-Scan transducer

Pulsed Doppler Transducer

The image produced by the B-Scan Unit is used to determine the coordinates for the Doppler transducer

Area of blood vessel investigated by the Doppler device

Figure 6.28 Duplex system

7

Computing

Computers and microprocessors are increasingly embedded in medical apparatus. Their use ranges from their appearance in data acquisition systems used in Intensive Therapy Units to Computer Based Patient Record systems used in primary care. We will examine some of the technical aspects of this use and look at the impact it will increasingly place on requirements for data security and archiving.

The desire to automate simple or repetitive tasks is natural. If technology is available which can undertake tasks in a reliable manner without tiring, the need for tea breaks or complaint, then why not use it? Automation clearly presents opportunities, and as technological development proceeds, the incorporation of microprocessors into equipment where previously there was specialised hardware becomes economically attractive.

This simplistic view can lead to problems. It may be attractive if you are a manufacturer of microprocessors to maximise the scope of their use. However, the ability to obtain and record information should not imply that it is either worthwhile or desirable. When patient records are obtained, it is desirable to retain the information for a period of perhaps several years to ensure that any need to reinterpret the data can be met subsequently. The tendency with paper records obtained manually is to seek only that information required on a fairly short term basis. Computers may afford the means to record vastly more data, such as continuously monitoring a patient's ECG and other physiological data. It is worth questioning how fine the detail of retained information should be.

Another aspect of computerised data handling relates to data security. Increasingly, and for good reasons, computers are connected to networks. Their connectivity means that there is a finite risk of the confidential data they hold on a patient being subjected to unauthorised access. This might come about either by deliberate and malicious attempts to access the data, or by inadvertent disclosure by an authorised computer user. Malicious access can largely be prevented by adequate system design: unfortunately many popular systems at present in use offer no data protection. The problem of inadvertent disclosure is potentially more difficult to avoid, since it requires understanding of the nature of both the information held and of any computer network's capability by users of private data.

7.1. Classification of Computers

Computers have traditionally been classified into different types, such as minicomputers, mainframes and Personal Computers (PCs). The distinction between these types is to a large degree blurred. Possibly a better differentiation relates to machine cost. The purpose of classification is to enable the description of the machines' performance in the broadest sense. The following sections of this chapter examine facilities which are appropriate to some machine types but not necessarily others.

For our purposes, we consider a mainframe computer as one which affords a wide range of facilities for concurrent data processing and management. It would be able readily to connect to networks probably without seriously compromising the machine's performance. The machine would be expected to be able to hold large databases and provide proper control of their access. This sort of machine would normally not be appropriate to undertake real time data acquisition. Most of all, mainframe machines are expected to be expensive: this may make their purchaser feel good.

Personal computers have for a number of years been based on microprocessors. The Intel architecture has dominated the market since it was adopted by IBM for their machines in the early 1980s. Rivals exist based on the Motorola 68000 series of microprocessor. If we disregard for now their relative merits, the striking shared characteristics are simple operation and cost effective processing. As the machines are targeted at the cheapest market sector, they are rather weak in the sort of facilities used by technical programmers. They are, however, viable for use by an individual, often costing little more than the hardware needed just to access a mainframe class computer. On their own, their data protection facilities range from non-existent to minimal, effectively precluding their use with highly confidential data. One should of course bear in mind that the security of a computer system is unlikely to be any better than the lock on the door which prevents it being stolen!

Somewhere in between these classifications reside minicomputers and personal workstations. If these may be separately identified, they offer more sophisticated facilities than PCs, particularly in aspects of data integrity and security. The software development facilities are typically much more advanced for reasons which should become apparent. Either minicomputers or PCs are likely to be found in roles involving real time data acquisition, where the data gathering function is likely to be handled by a dedicated microprocessor in most commercially developed apparatus.

7.2. Outline of Computer Architecture

The architectural description of a computer is used to define the machine's structure. It presents a rather different view of the machine from its simple classification. The description comprises definitions of the machine's interfaces at a number of points of interest. Taking a view of the whole system, it must encompass both the machine's hardware and software. The definition is likely to be modular, with each component taking in a different aspect of the machine of defining its interfaces with different degrees of complexity.

For example, from the viewpoint of a programmer who wishes to provide support for new interface hardware, the starting point would probably be the machine's instruction set. This defines the operations that the processor may perform under program control. This programmer will need to know technical details such as how the machine's address space is laid out and how to access device registers. If the new hardware is to be usable, it may need to be accessed by other programs, so a thorough description of interfacing the machine's input and output to the operating system is also required.

A different example would be of a programmer who required to define an application which was required to search for items in a database. This programmer would be concerned with the format and sequence of data exchange requests to ensure that the database was properly updated and could remain consistent.

Neither of these activities should in principle require a knowledge of the implementation of the underlying hardware and software. They both require details at an appropriate level of the machine's interfaces and control mechanisms. The following sections are intended to clarify the nature of this sort of description in a manner appropriate for the major applications of computers in medicine.

Another viewpoint in the classification of types of computer system uses a modular description of a machine. However, in this analysis we do not examine the machine's components: instead our task is to split it into a number of levels of abstraction or complexity. For instance, an electronics engineer may view a computer in terms of the logic gates from which it is constructed. A FORTRAN programmer would see little at that level, but may well know about the machine's overall performance and the usefulness of its program libraries. Another equally valid view of a system would be of its capacity to undertake the sort of transactions required to support a seat reservation system. These various viewpoints (the list is not exhaustive) regard computing systems in terms of 'levels of interpreter' (Figure 7.1) as they are task oriented, rather than specifically technology based.

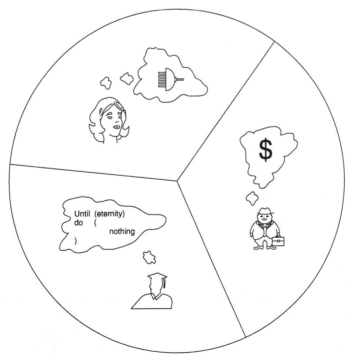

Figure 7.1 Levels of interpreter

7.2.1. Hardware

The hardware of a computer for our purposes comprises its electronic circuits. A typical structural overview of a simple computer, such as a PC or a minicomputer, is shown in Figure 7.2. In this level of description, all of the input output controllers and the processor are regarded as simple hardware functions. In fact in many cases they may themselves be constructed from components which themselves contain further microprocessors. The processor itself may be built from simple logic sub circuits which must be programmed in order to function with the computer's defined instruction set.

The components we have identified in Figure 7.2 are outlined in the following sections.

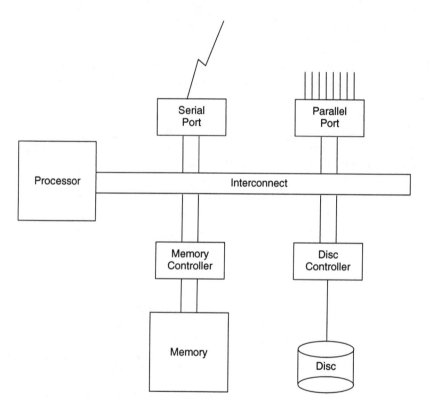

Figure 7.2 Outline of a computer system

7.2.1.1. Processor

The processor, or CPU, executes instructions. This is the central component of the computer which arbitrates and co-ordinates the operation of all other components. The instructions which the processor can undertake define its primary characteristics. They are received by the processor as binary encoded patterns in a sequence to control the operation of the machine. A programmer who requires to use the machine's instructions directly will normally do so through mnemonic representations of the instructions.

The complexity of function of computer instructions is something which has varied over the years owing to the changing benefits of the increasing density of electronic circuits which may be built. In other words, the degree of integration, or the complexity of circuit which could be laid down on a single silicon device has rapidly increased. However, optimal design strategies for a given level of technology do not always follow on from those of an earlier generation in a straightforward evolutionary manner as constraints are lifted. For instance, by the late 1960s, it was usual for a computer instruction set to contain a fairly wide range of operations which were aimed at simplifying the task of generating machine instructions from programs written in high level languages, such as FORTRAN. This approach to instruction set design leads to the need to decode separately each instruction, and imposes high costs for developing sufficient logic using small scale circuits. This problem may be overcome by the use of micro programmed machines which are in essence built from very simple, but fast and flexible logic machines with limited capabilities. They are enhanced to provide the intended level of architectural support by the use of a 'Micro program' in very fast access memory local to the processor. This sort of approach was commonly used in the computers developed in the decade from the mid 1970s. This approach is known as the Complex Instruction Set (CISC).

An alternative is to develop a computer without an internal micro program. Instructions are directly enacted by the machine's logic in a single machine cycle. This means that the area of the processor is used entirely for supporting its real instructions without the overhead of control and decoding logic for another level of architecture. This is known as Reduced Instruction Set (RISC) architecture. The machine's performance for single instructions is therefore optimised at the loss of sophistication of the instruction set. The result is that more work has to be undertaken by high level language compilers in optimising the code they generate: it nevertheless typically contains more instructions than CISC, but performance with the same generation of electronic technology is nowadays generally better.

7.2.1.2. Memory

The computer's memory is nowadays made from Random Access Memory chips. At the time of writing, most semiconductor manufacturers supply devices which can store a maximum of 4 Mbits of data which is addressable in eight bit units. This packing density has quadrupled every three years for the past 20 since silicon memories generally replaced magnetic core memories on computers.

The memory devices are accessed by the processor via its interconnect (see section 7.2.1.3). The amount of data read by the CPU in a single request depends on the nature of the machine. Typically however, computer designers ensure that data are moved in larger units than the program strictly requested, since most programs are highly localised in their memory accesses. The time taken to move eight bits of data would certainly be as long as that required to access the typical 32 bit width of common computer interconnects.

The memory chips are therefore interfaced to the interconnect via their own controller which is responsible for managing a degree of error checking and ensuring that data are presented to the computer in the correct sequence and observe the mechanisms required by the interconnect.

Most modern computers, including PCs, define their memory in several levels. The most expensive, and fastest memory is logically closest to the processor where it may be accessed with the minimum overhead. This memory is normally accessible by the computer in about one processor clock cycle. It is not normally therefore accessed via the interconnect. This is known as 'cache' memory which relies again on the localisation of both instructions and data. It is copied from the main memory when an access request cannot be directly satisfied: copying of the cache's contents back to the main memory is done differently on different machines. Typically programs read much more data (and instructions) than they write data (by about a factor of ten), so processing delays are not apparent when data are returned to main memory from the cache.

The time to access main memory is typically several times slower than that of the cache, but if the cache size is adequate for the application, then 'hit rates' (the probability of locating data in cache rather than having to look in main memory) of around 95% may be expected. This strategy both ensures that the cost of memory may be minimised by purchasing only a small quantity of the highest cost memory, but it also favourably impacts on the cost of the computer's interconnect which would otherwise have to carry much higher traffic between processor and memory.

7.2.1.3. Interconnect

The computer requires a pathway to enable data and control information to be transferred between its various components. The form of the pathway varies between simple busses on PCs and most minicomputers to larger switching structures with multiple paths and redundant structures on mainframes. Clearly the more sophisticated structures cost much more to implement.

Of primary interest here are bus interconnects. These provide the ability to send and receive data and addresses. They are normally multiplexed (i.e. shared by switching between activities) on a demand basis. A device which requires to send information to another on the bus must first obtain bus ownership from the master (normally interface electronics in the processor). When it has been granted ownership it is permitted to use the bus for a period. Some busses require that a device shall relinquish ownership after a predefined period to prevent a single device from dominating its bandwidth.

Having obtained the bus, the device then must indicate where the data are to be sent to or received from by asserting the address of the other device on the bus. Some busses have sufficient lines to permit this operation in parallel with data transfer, others do not. The responding device is then required to enact the read or write request and acknowledge via the bus protocol that it has done so.

Data may be sent to the processor via one of several mechanisms. In the simplest case, all major actions are taken by the processor which addresses the required device to examine its status. If it is ready for data transfer, then the processor may move the data it desires to the device's output register. If not, it may loop until the device is ready. This procedure is clearly wasteful of the processor's time if it is able to undertake another task at that time instead of waiting.

An alternative is to allow the device to provide notification of its status change to the processor as soon as it happens, and without the processor constantly having to monitor it through instructions. In this case, the device asserts an 'interrupt' line: the processor may then 'grant' the interrupt at a convenient point in its instruction stream. The device identifies itself typically if this mechanism is used by placing an address on the bus to tell the processor where to find an 'interrupt service routine'. The service routine is entered by the processor only once its previous state has been saved. This routine permits the processor to examine the device and undertake a few simple tasks such as noting the change of device state and moving the transferred data before it resumes its previous operation from its saved state.

Many machines permit a further level of sophistication which is necessary for devices which need to transfer large amounts of data to or from memory quickly and with a minimum overhead on the processor. These might be discs or network devices. The use of Autonomous Input and Output, or Direct Memory Access (DMA), permits that transfer of a block of data for each interrupt to the processor. The device which performs DMA must be more sophisticated than one which performs its operations entirely under processor control. It must be able to move data to or from the set of memory locations defined by the processor in sequence and must obtain bus bandwidth in the correct manner to avoid disturbance to the machine's overall performance. The processor having started a DMA operation, normally passes responsibility for its completion to the device. The device interrupts the processor on completion when it provides information again about its status.

7.2.1.4 Disc Drive

Computer discs provide bulk storage for data. Most current discs use magnetic media for storage of data as it provides a means for reasonably stable non-volatile data storage, at a high density. Most modern high density discs are built from a set of platters coated with a magnetic oxide film which rotate at high speed inside a dust free environment. Information is read from and written to the surface by recording heads mounted on a swivelling arm which permits the heads to fly at around 0.2 µm above the disc surface. (Compare this with the thickness of a human hair at around 40 µm.)

The size and number of the platters depends both on the age of the disc technology and its intended purpose. At the time of writing, the most advanced technology drives are being built for PCs. They have the highest data packing densities and occupy the smallest space. The leading edge producers for these are moving away from constructing disc platters of 3.5 inches in diameter to around 1.5 inches, and accompanying this with a change in platter material from aluminium to glass, owing to the latter's lower ductility.

In the more sophisticated market of mainframe discs where higher standards of availability are required, current discs are of 5.25 diameter platter, rotate at 5400 r.p.m. and store about 3.5 GB of data. Quoting figures like these quickly dates any book! However, the points in principle to note concern how data are accessed on the media and the causes of delay in locating data. The impact of disc access should be understood in regard to the later discussions about filing systems and databases.

A disc is normally divided into a number of 'sectors' which store data. Each sector contains both the user data block and information which is used only by the disc's control electronics to validate the data. Sectors are arranged in circular 'tracks' around the disc. A set of tracks vertically but not laterally separated is called a 'cylinder', so the number of tracks comprising a cylinder depends on the number of recording surfaces. Data are stored then in sectors which are themselves stored on tracks on a particular surface. Many computers now distance the central processor from this knowledge of the disc's layout by referring only to a 'logical' disc whose sectors are numbered in sequence from 0 to the disc capacity. The sector size is dependent on the machine's organisation, but is typically between 512 and 4096 bytes.

Access to data on disc when we know its location requires both rotation of the disc and movement of the recording head. The rotational movement, the 'latency', depends on the speed of rotation of the disc spindle: for a typical disc drive which rotates at 3600 r.p.m., this is 8.3 ms on average (half the rotational period). The time to move the head is controlled by the amount the head has to move, its inertial moment, and the dynamics of its actuating system. The time to settle on a track is therefore very varied, but current systems exhibit times of around 10 ms. Not surprisingly this time is reduced with reductions in disc dimensions, although the fine settling time is adversely affected by increasing storage density.

The result of these factors is that the performance of a disc in terms of its data rate is more controlled by its physical movement than by the inherent data rate which the recording head may sustain. Most operating systems support files which are discontiguous (see section 7.2.6). The result is that when reading a file, a number of movements will occur. Whilst the read/write rates supported by the disc head may be greater than 2 MBs^{-1}, few systems can achieve data rates of half that value in practice owing to file fragmentation.

The performance of a disc system may also be significantly affected by providing two forms of improvement in the control electronics. Firstly, we may take advantage of locality of data, whereby adjacent sectors frequently need to be read. The simplest method is to provide some local storage on the disc controller which makes it possible to read data on either side of the sector requested, and possibly the whole of a track. The data are retained by the controller until a request arrives for the additionally read sector, but may then be returned to the requester without the mechanical delays described above. Secondly, the movement of the disc head may be optimised to take advantage of the reduced seek time required when its movement is minimised. In this case the control electronics re-order read and write requests to optimise disc head movement. Any practical system which provides this speed up, which can improve throughput by around 30%, has to be sufficiently capable to prevent excessive deterioration in performance seen by requests for data away from areas of high request density.

The other commonly encountered disc technology uses flexible media. Since the mid 1970s a number of physical disc sizes have been used with increasing densities of storage, and significant performance and reliability improvements. A plastic disc is used which is coated with a magnetic oxide film: the read/write head contacts the surface of the disc. Clearly if the media may be exposed to dirt and temperature fluctuations, there is plenty of scope for misalignments and other errors, so the recording density is many orders of magnitude worse than that of a hard disc drive. To avoid excessive abrasion, the rotational rate of floppy disc drives is also kept small, and the resulting data rate is much smaller than that of a hard drive. However, floppy discs can be useful cheap data interchange media, even if their capacity remains irritatingly too low to be of use for other purposes.

7.2.1.5. Parallel Port

The description of the interconnect above should give some idea of the complications of using it directly for input and output. Ultimately, it is obviously the means by which data are moved to and from the computer's memory or disc. The complications of using it directly, and the chaos caused by its misuse mean that it is almost always necessary to provide separate interface devices which access the interconnect on behalf of the user. Furthermore a well specified interface may provide a user view of the computer which is convenient and can be reproduced when other underlying busses are employed.

The purpose of the parallel port on a computer is to provide this independence of the bus protocols and architecture. The parallel port provided on most PCs is intended normally to enable the connection of a printer. The port is therefore able to read and write data in eight bit units in parallel. It has control and status lines which make both the user and the computer aware of the port's status and enable flow control to be implemented.

7.2.1.6. Serial Port

Serial ports are frequently provided on computers to enable data transfer on a pair, or more often two pairs of wires. Data bits are transmitted and received in sequence either at a predefined rate or alongside a clocking signal. This form of communication originated in essence long before electronic computers with telegraphy. This section is intended to outline the mechanisms used in respect of the computer: a later section looks at more details regarding access to public data networks which use similar facilities.

The computer's serial port provides the facility to convert information presented in a parallel form on the Interconnect into the serial bit stream used externally. This form of connection is suitable when data must be transferred over longer distances than can readily be handled by the output stages of logic chips. As this is a frequent requirement a range of support devices is available to provide the required functionality.

Serial data are frequently transferred in characters, of typically eight bits in length, with self framing information provided by start and stop delimiters. This mechanism, known as asynchronous data transfer, is suitable when small amounts of data are to be transferred or when the data require immediate interpretation, such as when a computer uses keyboard input to update an editor's screen continuously.

An alternative scheme is to transfer messages of greater length (normally between 128 and 4096 bytes) without separation between the bytes. In this case, synchronous data transfer, the data frame is delimited with header and trailer patterns. The bit timing is defined by separate clock pulses which are exchanged between the two parties transferring data.

In either of the above cases, the connection of the serial port to the outside world is somewhat more complicated than it has been represented so far. The port's interface is presented as a number of connections which obey the rules of one of the standards such as RS-232 or RS-423. These define a number of 'circuits' each of which carries certain signals, such as received data, transmitted data and perhaps clocks. There are also circuits for exchanging information about the state of the interface which may be used for flow control or managing connections. The interface standards define the requirements for the signals' timing and voltage levels, hopefully to enable devices supplied by different manufacturers to communicate. Unfortunately some of the earlier standards in this area leave something to be desired as they have sufficient scope for interpretation that not all compliant devices will invariably connect. The topic of network connection and standards is covered in more detail in section 7.4.

7.2.2. Operating System

Computer hardware in itself is next to useless. To perform functions on behalf of its user it must have programs loaded and running. If it is to communicate the results of its work to the outside world, it must undertake input and output. The earliest computers had hardware which supported the minimum of in-built functionality for this communication. The program which undertook the initial loading of programs required to be entered manually via switches.

Frequently after the user program was loaded, it needed to contain all of the instructions required to control the hardware to generate output. Fairly early in their development, this sort of function was standardised for each developed machine. The repetitive tasks of ensuring that the machine's hardware was properly controlled was delegated to a supervisory program which had the responsibility for loading programs and scheduling work. This idea has progressed so that now Operating Systems provide many of the common functions used on computers avoiding the need for their constant redevelopment. Operating Systems themselves are to some extent standardised, so that the user application may be run on very different hardware without change as the System provides a sufficient interfacing layer to obscure the underlying distinctions.

Having taken on the role of isolating the user's applications from the underlying hardware mechanisms, the Operating System may be placed in a supervisory role where it is the guardian of the machine's resources. This has the effect of ensuring that the machine's hardware resources are effectively used and fairly shared between concurrently running applications. The System grants the user program ordered access to Input and Output devices, seeking to ensure that the requirements of different users do not conflict. It similarly allocates memory in a manner which should ensure that the operating system's own integrity is not compromised and that other applications' data are not inadvertently or maliciously accessed.

A goal of the operating system of a complex computer should be to ensure that the machine's resources are used in a manner which maximises their use by applications. This is the scheduling function which is discussed in more detail in section 7.2.5.

7.2.3. Input and Output Mechanisms

Control of input and output is one of the major functions of an operating system. It is essential that the system supervises this function for two crucial reasons. Firstly, it is the duty of the operating system to allocate the computer's memory resources. Since an input output device must be able to move data to any part of the machine's memory, the transfer must be undertaken by a component of the operating system. Secondly, and of equal importance, the provision of an interface to the input and output facilities of the computer by the operating system means that the details of data transfer may be largely obscured from the user. The transfer may be directed to a generic device, rather than to a specific location. This means that programs may be used in differing environments without modification to take account of trivial differences in configuration.

7.2.4. Protection

If the supervisory functions of a computer operating system are to be relied upon, then they must be able to control how the resources of the machine are accessed and allocated. User programs, if they are to behave properly need to able to verify that the data areas which they access are those intended.

A common scheme which provides the ability to support these functions is to implement in the computer's hardware a differentiation between two or more 'levels of access'. The inner mode, often known as 'kernel mode', is able to undertake the machine's full instruction set, and is normally defined by the operating system to have controlling access to the machine's memory and other resources. The outermost mode (frequently called the 'user mode') is restricted in the memory accesses it makes to those which were permitted by kernel mode code. It is also unable to undertake the hardware instructions which change the machine's state, and particularly those which directly alter the access mode.

By defining a hierarchy of access, the operating system may ensure that only permitted accesses are made to memory, and that those which were not are handled appropriately by the operating system.

This sort of scheme must be implemented on any usable multi user or multi tasking system, since otherwise the failure of one application is able to affect other applications detrimentally. It is not implemented on MS-DOS since the system architecture owes its origins to early microprocessor technology which did not contain adequate facilities for its support.

7.2.5. Scheduling

In multi tasking and multi user systems, the operating system is responsible for deciding how to order work. In these systems, separate tasks are normally referred to as processes. At any time a number of processes is likely to be waiting for the completion of an Input or Output request. This can either be because it is waiting for user response to a previous operation, or perhaps the relatively low speed of disc access (several 10s of milliseconds to access data) by comparison with the machine's instruction speed (perhaps around 10 million per second). The function of ensuring that both the processor and input/output resource utilisation are maximised is handled by the operating system's scheduler. Apart from the goal of seeking to maximise utilisation, a system will normally be designed to ensure that the variation in response time seen by different processes is minimised.

Several methods of work scheduling are in common use. The simplest allows a process to use the CPU either until it is blocked by the need to wait for input or output, or until it has used a quantum of processor time. When one of these conditions is met, the blocked task is placed at the end of the worklist, and the context used by the processor is switched to the next process in the list of eligible tasks. The time quantum is chosen to achieve a balance between the overhead of switching between tasks and ensuring that all tasks in the worklist receive some attention within a reasonable period.

This scheme may be enhanced by assigning priorities to the computer's processes. The machine will then activate the highest priority available task. In some systems, the priority is dynamically controlled. Processes receive a boost to their priority when certain events occur, such as the completion of input. Such a process is statistically most likely to require a small amount of service before again blocking through undertaking further input. This strategy therefore can help achieve increased input/output rates with fairly little detriment to the computational performance.

Scheduling of work in a computer also takes place at the level in the machine of interrupt service. This form of scheduling is largely controlled by hardware events which require time critical service to avoid data loss owing to a further event occurring which causes overwriting.

7.2.6 Filing Systems

Computer disc systems are intended to store large amounts of data. The data are normally organised into units of files which may contain sets of related items, such as a document (this chapter is one such), or an executable program. For a modern computer to be usable, there is likely to be a very large number of files stored on any given disc volume. To make the individual files reasonably accessible they are listed in directories by their name. Most systems permit directories to be defined within other directories (they are sub-directories). This overall strategy permits an ordered definition of the data held on the computer's filing system so that the data required by a user may readily be accessed via a logical path. This section describes in outline the organisation of one form of filing system, so that the use of individual files may subsequently be contrasted with the description of a database given in section 7.5.

Start by picturing a special file on a disc which is expected by the operating system to start at a particular block on the disc. (If the system is to be reasonably reliable it will probably keep a back up copy of this information at another location in case of localised disc failure.) The special file contains information about the location of every file held on the disc. The information points to the starting location of the files, and the length of information held from the starting location.

The first few blocks of our special file, called the 'volume index' themselves point to the files used by the system to control the disc volume. The two most important components are the description of the index itself and the pointer to the top level directory on the disc. Directories are special files which contain the names of all the files which users have grouped together, including the names of subdirectories. Together with each filename must be a pointer to the descriptive block within the index volume for the file being described.

We need one further aspect of information. In order to allocate blocks from the discs reasonably efficiently, the disc index points to a file called the bitmap which has one bit for each allocatable block on the volume. Bits are set to indicate that the corresponding block is occupied and cleared when the block is released.

As files in this sort of system come and go, the disc space fragments. Perhaps the first available group of blocks in the volume may be insufficient to accommodate a new file, or otherwise it could be extended at some later point in its lifetime. In either event, there is a possibility that some files will not allocate as a contiguous group of blocks. In this case, the index file needs to contain more than one pointer to a file extent: each of these file extent pointers clearly has to indicate the number of associated blocks starting from the first block in a group.

7.3. Data Acquisition

Computers are routinely used for data recording. They provide the flexibility to obtain and manage received data in a manner which permits its subsequent retrieval, examination and interpretation. The data may be obtained from physiological measurements or perhaps from image acquisition. In either case the data require to be transmitted to the computer in a manner in which it has been set up to expect. It ultimately must be sent to the computer in digital form.

The conversion of physiological data into a digital form was discussed previously in Chapter 4. In that chapter, we discussed several electronic means for conversion and the system requirements for data precision and sampling rate. When we specify a computer to carry out the data acquisition and processing functions, we must additionally bear in mind its performance constraints. Apart from figures which quote the simple bandwidth of the electronic components in the computer which receive the data, we must be aware of the rate at which the computer's processor may accept interrupts. Each data transaction involving moving a block of data to and from memory typically requires around 1000 machine code instructions. A machine which is capable of processing a million instructions per second (slow by modern standards) would therefore do nothing else but receive data if interrupts occurred 1000 times per second. However, even this transaction rate is well in advance of its real capacity, since when its data rate saturates, its service time for interrupts would progressively increase to a level which would not be sustainable. It would clearly also expect to undertake some other processing and to move the acquired data to secure backing store, further loading the processor. A machine offering that performance level would thus be doing little else if it required to receive more than 100 interrupts per second.

7.4. Computer Networks

A computer network enables connected computers to exchange information. As there is a wide variety of computing equipment on the market, equipment from different suppliers who employ various standards for their data must use common standards for communication if any data exchange is to be meaningful. Without this apparent chaos, there would be insufficient diversity of supply to promote technical development. Network standardisation, whilst it necessarily relies on using a lowest common denominator of the facilities available on the participating computers, is however enabling rather than restrictive in its nature.

Computer networks have developed rapidly from their inception in the late 1960s. Initially they provided facilities for the connection of numbers of the same sort of computer, thus

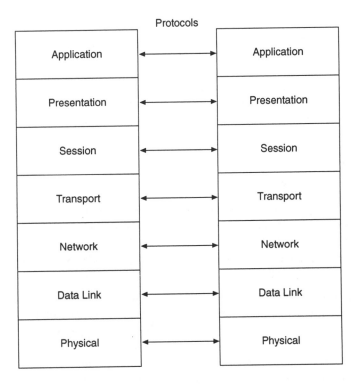

Figure 7.3 ISO network model

avoiding the need to convert between different data types and interpretation. This form of network was particularly useful in assisting the avoidance of excessive local loads. Problems came about with the development of new computer architectures and operating systems which did not readily fit into those networks. Heterogeneous networks followed a few years later and led to international collaboration in the definition firstly of a model for partitioning network function and subsequently the individual standards used for communication.

Modern networks provide the user the ability to share and exchange data with other users of the network, to send and receive electronic mail messages, and to access the facilities of remote computers. The model for communication was developed under the auspices of the International Standards Organisation (ISO), and published in 1978, as its seven layer model for open systems communication, shown in Figure 7.3. The layering defined by the model is intended to separate conceptual levels of functionality so that services can be logically grouped.

Each layer in the network model uses the services of the layer below it, and provides services to the one immediately above it. Corresponding levels on communicating computers are said to be in 'Virtual' communication with each other. This architecture means that there is a good degree of decoupling between layers so that the model is able to accommodate technical change in layers readily without the effects being felt outside the scope of the modified layer.

In an introductory text, the aspects of computer networks that are of primary interest relate to the technology and performance of the lowest layers of the network model which physically communicate with one another, and the high level facilities offered in the Application Layer of the model.

7.4.1. Low Level Protocols

Reference is frequently made to Local Area Networks, to distinguish them from Wide Area Networks. Apart from obvious differences in physical extent, the technological distinctions relate to the speed and reliability of communication. The designer of a local network should expect that at any time a significant number of the participating computers will require to communicate with each other. They should therefore be configured in a manner which makes communication straightforward. For instance, they should expect to have buffers available to receive any anticipated messages, and obviate the need for low level flow control. The network should be designed to have low error rates to avoid the need for the frequent exchange of supervisory messages.

Wide area networks on the other hand are likely to facilitate communication between machines whose contacts are sporadic. The connection bandwidth is typically much lower, so the conditions are set for needing flow control. The greater distances used increase the scope for noise to interfere with the data. There is therefore a much greater requirement for a low level error correction mechanism.

The networks described below are representative of their types. The discussion here is not intended to be exhaustive, but indicative of the levels of performance and relative cost of the different technologies. More detailed introductory texts are Black (1989) and Tanenbaum (1989).

7.4.1.1. X25 Networks

X25 is the name of the set of recommendations defined by the CCITT (Comité Consultatif International Télégraphique et Téléphonique) to enable communication via a public network. The definition arose from prototype networks which used serial data links and was promoted by the international telephone authorities to assist their participation in this market place. The low level of communication uses frames which carry messages, channel information to identify the particular information and a checksum to ensure reasonably reliable transfer of data. At the higher levels, X25 uses 'Virtual Connections' which support communication between participating processes on separate computers. The term 'Virtual Communication' is used to describe the circumstance when communication takes place between machines or services which do not have a dedicated electrical path connecting them.

Messages are routed through the network in a manner which is not visible to the user application. The protocol merely guarantees that information will be delivered in the order in which it was sent and within a specified time.

The speed of an X25 network is largely controlled by the data rate supported by the network lines used (within the performance limitations of the connected computers). Early public X25 networks used signalling rates of 480 or 9600 bit s^{-1}. More recent networks operate at speeds of up to around 2 Mbit s^{-1}.

7.4.1.2. Ethernet

The first version of Ethernet was developed by the Xerox Corporation in 1970. It provides for communication via a coaxial cable, with a signalling rate of 10 Mbit s^{-1}. All participating computers are connected to the network and are required to send information in 'datagram' frames which include the addresses of both sending and receiving computers. Receiving computers are required to accept only messages addressed to them.

A transmitting computer may send its data when it senses that the network is not in use. It must check that the information on the network corresponds with what it is sending to check whether another machine is attempting to transmit another message at the same time. It must clearly examine the network for a period corresponding to the maximum delay which a signal could experience in traversing the network. If its signal collides with a message originating from another computer, then the signal it sees on the network differs from that which it is currently attempting to transmit. It must then cease transmission: the other transmitter will also detect that its own signal is not appearing correctly on the network, and will also stop transmitting. Both intended senders then wait for randomly chosen intervals before sensing the network afresh to see if they can transmit their message. This strategy seeks to reduce the likelihood of a further collision once the transmission attempt is repeated.

The throughput of Ethernet may be expected to be up to around 60% line utilisation, representing around 600 KB s^{-1}. Between any two computers this figure is likely to be somewhat less owing to the performance limitations of the major components involved: the interface to the network, the processor and very probably the disc system. As a broad rule of thumb, maximum network performance is related to the instruction rate of the processor. Over many years of development of networks, and from the products of several manufacturers, a network throughput of 1 bit per second is obtainable from each instruction per second of processor performance.

7.4.1.3. Fibre Optic Networks – FDDI

Instead of using copper as a transmission medium, light signals may be sent along optical fibres. These afford higher transmission rates without making for significantly greater difficulties with impedance matching which are increasingly apparent with high frequency circuits.

The FDDI (Fibre Distributed Data Interface) standard defines a network which complies with the ISO model (shown in Figure 7.3) and achieves a 100 Mbit s^{-1} data signalling rate. The standard specifies that the network, which has a maximum circumference of 200 km, is connected as a dual concentric ring. The two rings permit data to rotate in counter rotating directions, and are preferably dually connected to each participating computer. This strategy is designed to provide for resilience of the network against persistent transmission failure as the network is able to reconfigure itself dynamically and automatically.

Access to the network is obtained by the use of a token. A computer which receives a special message, the token, is permitted to transmit data on the network. When it has finished transmitting, it must retransmit the token on the network, allowing another machine to transmit. Holders of the token are permitted to hold the token only for a time up to a configured limit. This ensures that the network's bandwidth may not be dominated by a single computer.

The current cost of installation of and connection to an FDDI network exceeds ten times that of Ethernet. This reduces the scope of its application to areas such as interconnecting hubs between other local networks. The cost penalty may be expected to reduce if fibre networks become more widely accepted and can take advantage of scale economies in the manufacture of their components. A serious present difficulty is in producing interface components which are capable of keeping step with the data signalling rate employed by the network to monitor the presentation of addresses and resignal received data.

7.4.2. Application Protocols

The Application Level protocols provide services which are accessed directly by users. They make use of the network via the services provided by the Presentation Layer which enables application services to define the format of messages to be exchanged via the network so that they may be mutually interpreted. The arrival and commercial acceptance of these standards has been a long time coming. From the initial publication of the Reference Model in 1978 agreement on the file transfer standard took until 1987 and Electronic Mail 1988. Of necessity comprehensive implementations and use take some further time.

7.4.2.1. File Transfer

The ISO file transfer service is known as the FTAM (File Transfer, Access and Management) protocol. It defines a rich range of facilities to enable the transfer of information between computer filestores. The core of its mechanism is its definition of an idealised filestore model. Each party to a filestore must represent exchanged data using Presentation Layer services in terms of this model. The protocol, in concert with lower layer network mechanisms enables files to be located and securely copied through the network.

The purpose of this service is to facilitate communication between machines which employ very different conventions for the storage of their data. The problem which it does not address is how to define the meaning associated with data exchanged between networked computers. The transfer of text information is straightforward and was possible with the forerunners of FTAM: the requirement for interpreting the transferred information does not stretch the ability of different systems. FTAM adequately addresses the issues of how to move data between computers and preserve its semantic content. However the problem remains that other applications, such as image storage, may not be standardised sufficiently between manufacturers to make this sort of information exchange worthwhile.

7.4.2.2. Electronic Mail

Electronic messaging via computer networks is becoming increasingly popular. The use of computer mail has arrived in a standardised form rather late in the day as there has been a proliferation of mail mechanisms from a variety of sources pushed on by a perceived need in advance of the completion of internationally agreed standards.

The ISO standard is known as MOTIS (Message Oriented Text Interchange System), which was derived from the earlier CCITT X.400 mailing standard. The development of the two standards was merged in 1988 and made properly compliant with the ISO network Reference Model.

MOTIS permits a user to send a message either to another user or group of users with the assurance that the message will be delivered. There are mechanisms which are designed to ensure that a message which fails to be correctly transmitted to its end point is not lost, but the fact is reported to its originator.

Electronic mail, whilst requiring very high implementation costs, provides an excellent means for communicating messages between people who would otherwise need to meet or telephone. The messages may be recorded and take advantage of being computer generated and stored. Electronic mail may then be useful in transferring results obtained via computer data logging, or perhaps images with a low probability of transcription error.

Delivery of electronic mail is assisted by the use of an electronic directory service whose function includes building lists of known names for users rather than cryptic usernames beloved of computer installations. This service has further benefits in terms of assistance in supporting mailing lists to groups of users and validating the source of messages so that the mail network's security and integrity may be guaranteed.

7.5. Databases

Databases are simply a computerised method of storing and retrieving data. Typically they permit the creation, modification and deletion of data along with facilities for viewing the data in many different formats. A database differs from a program with embedded data in that the data is independent of the program which accesses and manipulates the data. This independence provides many benefits as a number of different programs that use the data can be written and modified without interfering with other users. Additional indices are commonly added to a database to improve access time. Very primitive systems use simple files for storing data; however, this form of storage does not provide a means of defining relationships between sets of data. Consequently a variety of different types of database have been developed which address the problem of forming such associations.

In developing Database Management Systems (DBMS) the ANSI/SPARC 3 level architecture has proved to be a particularly useful model. In this model three levels are described. The lowest level is the internal schema which dictates how the data are physically stored. Naturally the programmer does not wish to get involved with such implementation detail, so the next level is described by a 'conceptual schema'. The conceptual schema provides a logical description of all the data stored in the database. Finally there is the external schema which is the interface the user will see, where the data are carefully presented and limited to that required by the user. Each level is appropriate to different users of the database and each level has its own Data Description Language (DDL) and Data Manipulation Language (DML).

7.5.1. Why use Databases in Medicine?

A clinician receives large quantities of information over very short periods of time. Consequently effective means of storing and updating data are required. These data may be stored in a database for a number of different reasons:

- Medical History
- Medical Summary
- Data Integrity
- Financial Audit

- Medical Audit
- Access time
- Data Security

The most obvious reason for storing data is to provide a medical history so that new decisions can be made in the light of previous information. Given the problems of storing and accessing very large quantities of data this reason alone suggests computerised systems would be advantageous. However, the medical profession is constantly seeking to improve care so 'medical audits' are performed. Such an audit necessitates reviewing patients with particular complaints and assessing the effectiveness of the treatment given. Gathering a statistically significant quantity of data for such groups is an immense task as the clinician has to wade through the paper based records held in many hospitals. By using a database sorting and reproducing information relevant to the audit can be done in minutes rather than weeks. Furthermore a database can produce a rudimentary report which summarises the information stored within it, thus improving productivity.

A related problem is that of accessing records for day-to-day usage. Records may have to be transported from ward to ward and hospital to hospital. Once in an electronic form properly implemented local or wide area networks make data transfer an easy task. On a rather more mundane level stand-alone machines, for instance in a GP practice, allow both access to records and printing of standardised forms. Prescriptions are an obvious example of where clear, unambiguous forms are required. As consultation times drop below ten minutes per patient small savings in time, due to automated procedures, become significant.

Duplicated information is another danger waiting to catch the unwary. One way of dealing with the problem of waiting for records is to produce a new set of records. However, this causes the duplication of information which at best is inefficient and at worst permits people to update one set of records but not the second set. Such activities cause inconsistencies in the data stored unless meticulous attention is paid to monitoring all the records on a patient. Computer systems offer a means of ensuring that duplication is minimised if not eliminated. Data integrity refers to the problem of ensuring that data is consistent across a database.

Assuming that a system has been devised which allows efficient and fast access to any patient's records, attention must now be paid to how those records are made secure. Arguably an open filing cabinet, or set of filing cabinets, containing the hospital records on a ward would provide fast access to patient records. However, such a system does not provide an effective means of ensuring security. A properly implemented multi-level computer security system can control access to records and log who has accessed which records and when.

Finally financial audit must be mentioned. Depending on government policy clinicians may be paid by the patient or by a set of regulations on payment for goals achieved. In either case detailed reports are quickly produced from a computerised system. When government policy is in a state of flux the ability to produce reports from a database that were not relevant in the past, but are now required, becomes imperative. The paper based alternative is to work through all previous medical records one by one.

7.5.2. Database Architecture

7.5.2.1. Flat File

The simplest way of storing data is simply to place those data in a file and search the file when the information is required. This method does not, in itself, hold information on how the data are structured. For example if a patient has a patient number, name, diagnosis number and diagnosis description then a single file could hold a list of such patients. Should the diagnosis number for a particular condition change, then all the patient records with that diagnosis number must be modified. Alternatively two files could be kept one with patient number, name and diagnosis number, the second with diagnosis number and description. This avoids the modification problem and will reduce duplicated information by storing the relationship between diagnosis number and diagnosis description only once. If this method is used then the program must be carefully designed to manipulate correctly the files so that the structure of the information is maintained. Other problems will then arise if more fields are added to the file as the original program may not be designed to cope with such additions.

7.5.2.2. Hierarchical Databases

Hierarchical databases impose some structure on the data; in particular a link is defined which connects one datum to multiple data. This type of structure is termed a parent/child relationship. The example given below shows a consultant, the 'parent', linked to patients and to medical teams, 'children'. So there are two parent/child relationships.

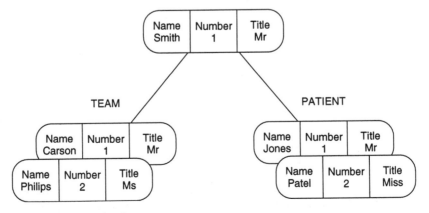

Figure 7.4 Hierarchical database

This form of structure is very effective, since each high level record has links with many lower level records. Problems arise when structures are not strictly hierarchical, for example if a patient is looked after by many consultants or if members of a team work for more than one consultant. Such relationships require a more complex structure.

7.5.2.3. Network Databases

A network database permits a more general structure than the hierarchical database. In this case a 'parent' can have many 'children' and a 'child' can have zero or more parents. Consequently a network of connections is created.

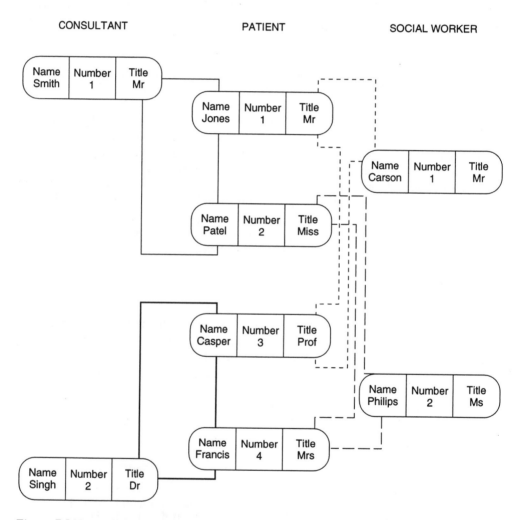

Figure 7.5 Network database

Such a database can be quite complex and hard to visualise, largely due to the existence of both data and connections. These connections (links) are also present in the hierarchical database but not in such a general form.

7.5.2.4. Relational Databases

Relational databases are based on Set Theory. Relations can be likened to tables. Each table stores information, the columns represent attributes and the rows a particular record called a tuple. The ordering of rows and columns is not specified, each row must be uniquely identifiable and each column has an identifier. Consequently any data item, called an atom, can be extracted by specifying the tuple and attribute. Extra attributes can be added later as the ordering of columns is not specified. Additional columns do not affect use of the database, except with regard to space consumed. Structure is imposed on the system by matching attributes in one table with attributes in another thus defining a link. The unique identifier specifying a row is known as the primary key of that relation, whereas an attribute that refers to data held in another table is known as a foreign key. Consequently both data and links are stored in one format, the relation. The following three relations Patient, Consultant and Social Worker illustrate how a relational database is represented. Each table has a set of attributes consisting of Title, Name, Number and additional attributes, where required, to link the relations. In this case Patient has the attributes CNo and SNo to indicate which Consultant and Social Worker is involved with a particular patient.

Consultant		
CTitle	CName	CNo
Mr	Smith	1
Dr	Singh	2

Patient				
CNo	Title	No	Name	SNo
1	Miss	2	Patel	2
1	Mr	1	Jones	1
2	Prof	3	Casper	1
2	Mrs	4	Francis	2

Social Worker		
SNo	STitle	SName
1	Mr	Carson
2	Ms	Philips

Figure 7.6 Relational database

7.5.2.5. Distributed Databases

There are a number of different ways of forming a database over a network. The following diagrams detail the notation used:

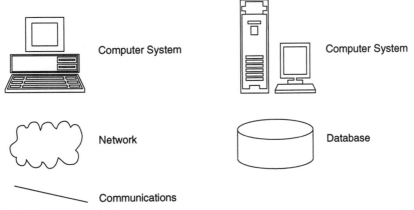

Computer System

Computer System

Network

Database

Communications

Figure 7.7 Network symbols

Centralised database

The centralised DBMS is placed on one system which users log onto either locally or remotely. This type of system is typical of mainframes where cheap terminals are used and specialist staff control the main computer. See Figure 7.8

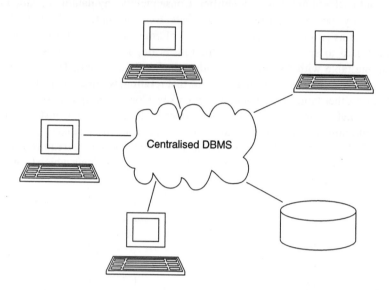

Figure 7.8 Centralised database

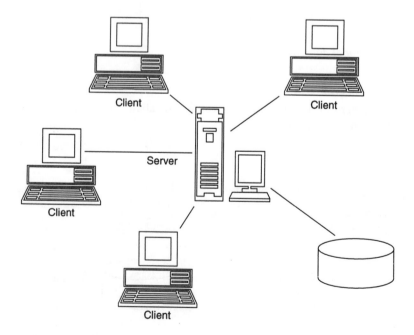

Figure 7.9 Client/Server architecture

Client/Server architecture

Control of the database is centralised on the server; however, the user interface is processed by the client computer. Client/Server architectures are of particular interest as they permit the use of cheap 'intelligent terminals' for local processing (Figure 7.9). The key advantage is that complex graphical interfaces can be controlled by the local processor as it is a local task whereas the database is controlled centrally by the server. This configuration is conducive to data integrity.

Distributed Database

The distributed database is held on two or more systems over a network; however, it is all part of one logical database. Such databases are particularly useful for multiple site usage; for example if a number of hospitals are linked forming a complete database, with each hospital holding local patients on its local database. As hospitals normally only access the records of local patients there is very little network traffic between hospitals. The main disadvantage of such a system is that duplicate data are often stored so that each site does not have to access the other site on a frequent basis. This duplication leads to data integrity problems if a site does not respond properly.

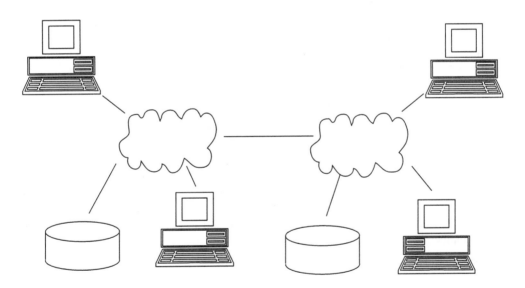

Figure 7.10 Distributed database

Federated database

At present there are many different databases available and clinicians would like to be able to access all of them with a common interface without worrying about where they are. The federated database is a collection of inhomogenous databases under different operating systems with a global database manger controlling the overall system. This collection of databases should appear to be one database to the user. There are considerable technical difficulties which frequently lead to facilities degenerating to the lowest common denominator. Careful costing of maintenance problems and interfacing software is advisable before assuming that existing disparate systems can be combined, as opposed to creating a new all encompassing distributed database designed for the task. See Figure 7.11.

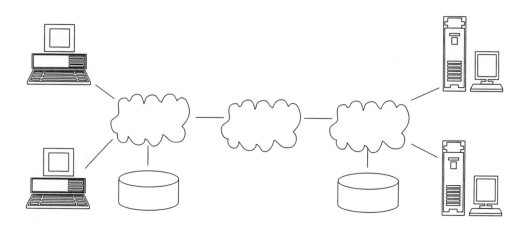

Figure 7.11 Federated database

7.5.3. Database Design

Databases must be carefully designed to ensure that data are not duplicated and can be efficiently accessed. As databases, particularly relational databases, are now becoming increasingly common an understanding of how they are designed is important. The following three subsections detail some of the key tools used to design a database.

7.5.3.1. Functional Dependency

Functional dependency is a key concept in database design. The underlying idea is that a record should only store data that are related to one concept. If multiple relations exist within a record then duplication of information is very likely. In order to facilitate design functional dependency diagrams are used to show which attributes are functionally dependent on other attributes. An example of this would be a relation PATIENT where the attributes are name, number and consultant. In this case the patient number is unique and therefore specifies a particular consultant, for that patient. Consultant is said to be functionally dependent on patient number.

PATIENT.number → PATIENT.Consultant

Patient number *functionally determines* Consultant

A functional dependency diagram of the consultant, patient, social worker database depicted in the relational database section would be as shown in Figure 7.12.

Note that each relation has its own functional dependency diagram.

In any relation an attribute that can be used as the unique identifier for that relation is known as a candidate key. One of the candidate keys will be chosen as the primary key. A well-designed database will ensure that the only functional dependencies will be direct functional dependencies on a candidate key (all arrows will go only from candidate keys to other attributes).

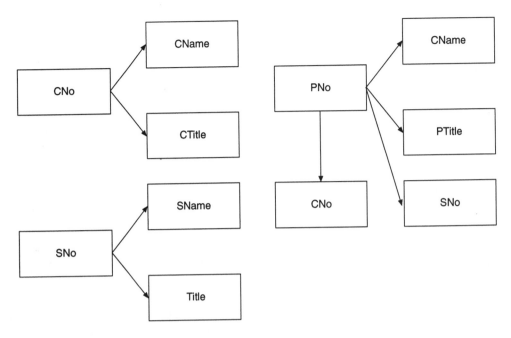

Figure 7.12 Functional dependency diagram

7.5.3.2. Normalisation of Relational Databases

A database can be expressed in a standardised form. These standard forms, termed normal forms, are designed to alleviate problems that may arise within a database. The process of converting a database to a normal form is known as normalisation. First normal form requires that each attribute of a tuple should be atomic. A typical situation is the patient/consultant problem.

Unnormalised

Consultant	Patient
Jones	Patel
	Smith
Hayward	Gray

First Normal Form

Consultant	Patient
Jones	Patel
Jones	Smith
Hayward	Gray

Patient in the unnormalised form holds two names in one tuple, this is forbidden in the normalised case.

Second normal form states that the database is in first normal form and that all attributes that are not candidate keys will depend fully on the primary key.

First Normal Form

Consultant	Vegetable	Patient
Jones	Cauliflower	Patel
Jones	Potatoes	Smith
Hayward	Cabbage	Gray

Second Normal Form

Consultant	Patient
Jones	Patel
Jones	Smith
Hayward	Gray

Vegetable
Cauliflower
Potatoes
Cabbage

As names of vegetables have nothing to do with the primary key (Patient) they should be listed in a separate table (relation).

Third normal form is the highest normal form commonly used during database design. For our purposes third normal form (3NF) will be treated as if it is identical to Boyce Codd Normal Form (BCNF). Strictly speaking the BCNF and 3NF are not equivalent but the difference is very small. Third normal form states that the database is in second normal form and that non-key attributes are not transitively dependent on the primary key. An alternate way of expressing this, BCNF, is that every determinant must be a candidate key. A determinant is an attribute which determines the value of another attribute.

Second Normal Form

Consultant	Age of Consultant	Patient
Jones	26	Patel
Jones	26	Smith
Hayward	20	Gray

Third Normal Form

Consultant	Patient
Jones	Patel
Smithers	Smith
Hayward	Gray

Consultant	Age of Consultant
Jones	26
Hayward	20

In this case although the Patient uniquely determines the age of the consultant, it only does so because Patient determines the Consultant and Consultant determines the Consultant's age. This relationship is known as a transitive dependency.

7.5.3.3. Entity Relationship Diagrams

Normalisation and functional dependency diagrams start from listing the attributes and producing appropriate diagrams, a bottom up approach. Entity relationship diagrams start by explicitly illustrating the relationship between entities. An entity is a self-contained object. By using this approach the designer can draft a design using a top down approach. Entities and relationships are drawn first and attributes may be added later.

Square boxes represent entities which are relations in a relational database. Diamonds represent relationships. Relationships are further specified by using **1** and **M**. So where one consultant has many patients a **1** is placed next to Consultant and **M** is placed next to Patient. According to this notation **M** means zero or more. Optional and mandatory relationships are also denoted by using O and I next to the appropriate entity/relationship link. Clearly a patient

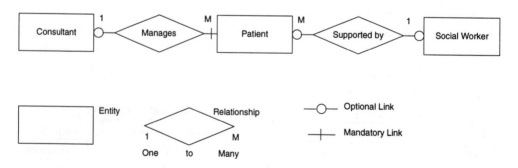

Figure 7.13 Entity relationship diagram

may or may not have a social worker, consequently the relationship is optional, denoted by an **O**. However, it is mandatory that a patient has a consultant; this is denoted by a l. Attributes can also be placed on the diagram as shown in Figure 7.14.

The primary key will usually be underlined if attributes are placed on an entity relationship diagram.

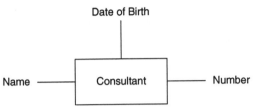

Figure 7.14 Entity with attributes

7.5.4. Medical Databases

General practice databases have become very widespread in the last decade. With the advent of cheap microcomputers medical records can be stored and accessed cheaply and effectively. Basic records can be kept on such systems such as storing patient details and a simple history. As cheap and readily available databases increase in size, images such as X-ray prints, may also be stored.

Typical services provided by medical databases are :

- Individual and family registration details
- Appointments
- Patient recall & preventative health screening
- Medical history summaries
- Prescriptions
- Formulary
- Practice reports
- Financial audit reports
- Searches on data for medical audit
- Links to hospital laboratories & databases

Spann 1990 and Rodnick 1990 discuss the merits of computers in family practice in the USA. Interestingly Rodnick suggests that a computerised system slows the GP down because of the time taken to enter records. Despite this statement computers are used in Britain to quickly produce clear prescriptions and automatically generate medical records along with statistics. Clearly the way software is implemented for ease and speed of use will strongly affect the usage of databases.

Hospitals cover a much greater variety of problems. One of the most significant areas is medical audit. Various databases are now used which store information primarily for audit or research purposes. Typically an historical record is kept listing some or all of the following: signs, symptoms, prognosis, procedures, findings and diagnosis. For example Ellis B.W et al 1987 discuss the use of databases in surgical audit, McCollum P.T et al 1990 apply a database to vascular surgery. Naturally a database is only as good as the data which are stored within it

Barrie (Barrie 1992) looks at how the quality of a database is compromised due to inaccurate or incomplete data. Feedback on the use of the database is emphasised as a means of ensuring the database is consistent.

Databases are not solely used for medical audit, nursing staff also use these systems for assisting them in 'care-planning' (Hoy 1990). The key advantage here is that a clear unambiguous plan is stored and becomes readily available to the data user. Information flow rather than statistics is the major factor in its use. Real-time systems are also used to collect and summarise data. Such systems transport data to a convenient point and convert it into a readable format. Examples of this can be found in Intensive Care (Fumai et al 1991) and Neurophysiology (Krieger 1991).

Financial services must also be considered as governments or insurance companies may require information from the hospital or general practitioner in order to make payments. Annis et al 1989 details how a database can be used to assist the finance section of the hospital, thus allowing money to stay with the patient rather than be directed to support services.

As databases continue to expand and images are stored and shared, attention must be paid to the type of networks that will be used and storage requirements. Allen (Allen et al 1992) discusses the problems of such networks and how to transport and store the large quantities of data involved.

The aforementioned uses of a medical database are by no means exhaustive; whatever system is required it is well to consider that a poorly implemented system is worse than no system at all. A database has an advantage over existing paper records because clear, accurate, up-to-date, reliable data are available, if this is not true then the database becomes expensive and useless. Great care must therefore be taken to specify what is required (Olagunju 1989).

7.6. Clinical Expert Systems

Expert systems are gradually being introduced into medicine and consequently clinicians need to be aware of the characteristics of this new technology. If the computer can successfully provide the information normally given by an expert then it deserves the title 'expert system'. This section looks at how the clinician approaches diagnosis and how the computer attempts to mimic this behaviour. Knowledge representation is a particularly important issue for these systems and a description is given of the three key methods used today. Finally an outline is given of the practical considerations associated with introducing an expert system.

7.6.1. Medical Reasoning

There are four main approaches to diagnosis (Williams 1982) none of which are used exclusively. They are exhaustive diagnosis, gestalt, algorithmic and hypothetico-deductive. An exhaustive approach covers all possible contingencies: in practice this is impractical as the number of unlikely but possible causes can be very large. In addition the tests required would be expensive and the time taken to check all possibilities prohibitive. The gestalt approach is one where an impression is received by the clinician from the combination of all the data at once rather than concentrating on any one aspect. Typically the demeanour, pallor and circumstance of the patient combined with the observed symptoms may as a whole be associated with a particular condition. Alternatively the algorithmic approach ignores the overall picture and concentrates on answers to particular queries. This method is a 'flow-chart' approach (Armstrong et al 1992) to diagnosing a problem, which allows little insight. Lastly the hypothetico-deductive approach is based on drawing together a number of hypotheses which are then proved or disproved by confirmatory tests.

All these approaches are used in normal medical diagnosis, with the exact combination depending on the experience and temperament of the clinician. During diagnosis reasoning would start with asking the patient questions and performing a clinical examination. Normally four or five active hypotheses will be considered during this time, possibly more amongst less experienced clinicians. Observations are then interpreted in the light of the active hypotheses to determine which, if any, can be reconciled with the patient's condition. Further tests then allow the clinician to identify the particular problem. It should be noted that the original hypotheses may have been determined by an algorithmic or gestalt approach.

Transferring this knowledge to an expert system poses a number of problems. The probability of a sign indicating a particular problem may not be known except in very simple cases. In many cases a combination of problems exists, particularly in the elderly, so assigning accurate probabilities is impossible. Even if accurate probability factors exist, the clinician may not know all the information required, particularly if a sign is difficult to detect. Assuming that the appropriate facts and probabilities are known there is no guarantee that the experts who programmed the computer have incorporated all the rules that they know let alone all those that exist. Finally the model of the human body on which the rules are based may be inaccurate, incomplete or both, either due to lack of information or disagreement amongst the experts. Despite these problems a body of knowledge does exist and clinicians do diagnose illnesses, so an expert system should be able to capture some of this expertise.

7.6.2. Expert Systems

Expert systems are intended to supplement the role of an expert giving advice. For example if a GP requires information on a specialist area he may approach a consultant who, in this case, is the expert. Typically an expert system should not only answer a problem but also give reasons and be able to demonstrate how the conclusion was reached. Naturally the questions asked by the system should follow an intelligent line of enquiry, otherwise the expert system degenerates into an exhaustive search of a database. If such a system is to be of benefit its interface must comply with its users' normal linguistic conventions and it should also be able to deal with symbolic concepts. Understanding and acting on concepts fulfils one of the criteria for Artificial Intelligence (AI). An expert system is one form of AI according to this definition.

There are many possible roles for these systems in medicine, diagnosis, therapy, financial audit, medical audit, teaching, research, and biological and medical engineering. As diagnostic aids they can assist the clinician in identifying the patient's condition, particularly where unusual problems occur (King 1990). As a therapeutic aid the system can give appropriate advice taking into consideration interactions between drugs, sensitivities of the patient and the latest information. Financial and medical audits have been covered in section 7.5.4 and are equally applicable to expert systems. Such systems can be given hypothetical cases or can be queried about real cases, in either event this can be very useful for teaching a student (Ferreira 1990). Once a large quantity of data has been acquired new hypotheses can be tested and conclusions drawn. This provides one obvious approach to supporting research in medicine.

Finally in biological and medical engineering a system which can automatically acquire and act upon data received from sensors attached to a patient can potentially enhance the power of existing monitoring systems. A variety of problems must be overcome before these systems can be realised.

When developing an expert system a few key areas must be addressed, namely how the knowledge is acquired and represented, how knowledge can be inferred from known information, how the information can be combined in a 'fuzzy' world and how the results can be expressed efficiently and effectively to the clinician. The remainder of this section on medical expert systems looks at knowledge representation as a way of characterising the systems currently available.

Three main types of knowledge representation are commonly used; production systems, semantic networks and frames. These representations are not mutually exclusive and will often be combined with each other and a normal database. Work is continuing on how best to combine various models (Ramoni et al 1992)

7.6.2.1. Production Systems

A production system combines rules and facts with a knowledge processor. By matching rules and facts new facts are produced which can then be matched with further rules. The section of the system that deals with processing this information is known as the inference engine. Typically the 'inference engine' will match rules and facts and then decide which has the highest priority and consequently which shall be triggered first. The rules are expressed as Boolean logic statements (i.e. conjunction, disjunction, negation, implication). Two methods can be used to find a solution in this system, forward chaining and backward chaining. Forward chaining starts from the facts and produces new facts by combining with rules. This process continues until all possible solutions have been found. Alternatively backward chaining starts with all rules (or facts) which can produce the solution. If some of the facts within a rule are unknown then this becomes a sub-goal. Backward chaining is generally preferred as the search space is normally much smaller than that searched in forward chaining.

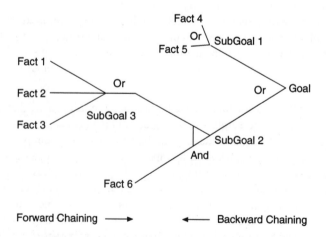

Figure 7.15 Logic chaining

Perfect rules rarely exist in the real world so uncertainty factors can be associated with a rule or a fact (Henkind 1988). These rules can be combined using numerical or pseudo-numerical techniques. Bayes' theorem forms a strong basis for a production system however the pseudo-numerical techniques are more common, the most well known of which was developed for MYCIN (Shortcliffe 1976). If the computer were infinitely fast with infinite storage capacity an exhaustive search of all possibilities would be feasible, however this is not so in practice. Two other approaches are used, namely activation criteria and metalevel rules. Activation criteria define a set of conditions which the rules and facts must fulfil if they are to be considered, whereas metalevel rules dictate how other rules are used. By using one or both of these methods the production system may produce a reasonable solution without the time penalty of performing an exhaustive search.

7.6.2.2. Semantic Networks

Semantic networks store knowledge by describing binary relations between entities, so 'x is a part of y' and 'the femur is attached to the tibia' are shown in Figure 7.16.

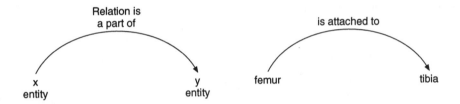

Figure 7.16 Semantic network

By adding more relations (oriented arcs) and entities (nodes) a graph is produced which is known as a semantic network. These networks form a simple and well-defined method of storing knowledge. Consequently it is an excellent way of representing knowledge. Unlike the production system, data acquisition is separated from the diagnosis in a semantic network. No diagnostic hypothesis is assumed, instead the system accepts or rejects states of the network. Consequently observations validate pathophysiological conditions which validate possible diseases which validate the treatment. Confidence factors can be placed on the relations so that a certainty factor for the results can be determined.

7.6.2.3. Frames

Frames store knowledge within data structures, each element of the data structure is called an attribute, property or slot. Slot is the normal term in AI applications. These slots hold information including other frames if appropriate. Each slot can have a default value or a value inherited from the slot of another frame. Procedures can be called as required or be automatically triggered by a particular event. These triggered procedures are known as 'demons'. An example of frame based storage of information is given in Figure 7.17 where two frames are depicted, Patient and Ward.

A brief review of intelligent systems has been written by Kulikowski 1988. Further general information can be found in Feischi 1984, Barr 1986 and Rich 1991.

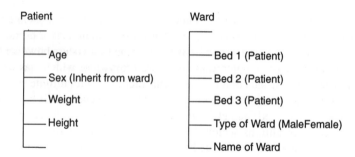

Figure 7.17 Frames

7.7. Privacy, Data Protection and Security

As information systems become an integral part of the working environment an appreciation of the problems posed by security, in various forms, is required (Lane 1985, Denning 1982). Physical security refers to damage to the medium on which information is stored. Data security covers the security problems associated with transferring, modifying or deleting data. In addition data integrity and privacy are two key problems. Data integrity involves the consistency of the data; privacy deals with an individual's rights.

7.7.1. Physical Security

Physical security covers both natural disasters such as fires, earthquakes and man made ones. Man made problems can be further categorised into accidents and deliberate damage. Accidents may be caused by incompetence, negligence or 'just' curiosity. However, more serious problems occur when damage is deliberate or data are stolen. A variety of methods can be used to safeguard data, the simplest of which involves having a safe area for the computer. Fire detectors and appropriate extinguishers are a straightforward method for dealing with one of the most common natural disasters. Training and well-designed systems alleviate many of the accidental man made problems. Finally any system must have an efficient mechanism for keeping backup copies, preferably off site. Taking these simple precautions provides a high degree of physical security.

7.7.2. Data Security

Data security covers areas such as unauthorised searches, information transfer, inference of private information, deletion or modification of data and unauthorised access using false identification. A simple expedient is to have a physical lock on the computer and on the room. Locked cabinets can store discs or tapes to prevent access to data as well as providing some physical security. Assuming the user has access to the system then the system should be able to limit how the data are transferred. If the user can transfer data to an unprotected file without any record of the transaction then it is effectively useless at preventing the authorised user from abusing the data. A more subtle form of abuse involves statistical databases which are designed to give general statistical information, not personal information. By putting careful conditions on a request it may be possible to identify personal information, for example 'How many 24 year old men, admitted on 13/6/91 to a medical ward in Cardiff, blood type A+ve have AIDS?'. This method of abuse is accomplished by inferring information, consequently inference control may be necessary.

In order to limit information to only the appropriate people they must be identified. Usually this is done using a password system although other techniques are available, such as security cards like the banking cash card combined with a Personal Identification Number. In some cases sensitive data must pass through channels which are not secure. Encryption can be used to protect such data although this doesn't prevent deletion. Modification can be detected by using standard data error detection methods such as Cyclic Redundancy Checking.

7.7.3. Data Integrity

Data integrity refers to the consistency of a set of data. There are two types of integrity: 'entity integrity' and 'referential integrity'. Entity integrity means that each data record is uniquely identifiable. Referential integrity states that a referenced record must exist. In practice a system should ensure that no conflicting data are stored, all data are up-to-date and deletions and modifications are automatically dealt with in such a way as to maintain the integrity of the system. Some modern database management systems are designed to cope with integrity constraints; however, not all do.

7.7.4. Implementation Considerations

This section summarises some practical considerations for the user. Obviously the safeguards listed earlier should be checked and if relevant implemented. If a password system is used, the system manager should decide if the password should be updated regularly with meaningless passwords, or if a permanent, easy to remember, password should be used. The former method can provide a high degree of security but if the user continually forgets the password or ends up writing it on the computer then a simple password is more effective. The 'choice of password' problem highlights another issue, namely that the system should be easy to use. If a system is awkward then people will either refuse to use it or will find ways to bypass the system. A typical example of the ease of use problem is the practical outcome of using a single level security system as opposed to a multilevel security system. For instance if a nurse must collect information from a system for which a consultant must provide the password, inevitably the nurse will discover the password as the consultant will not be willing to collect data for nurses or stand over them as they query the system. Backups are frequently highlighted as an important method of safeguarding data and yet are rarely used properly on small systems. This problem is caused by inefficient or time consuming backup procedures: if the data are important a satisfactory backup system is necessary. Finally the clinician must decide who will require the data and where. If a large number of users may eventually want to use the system on different sites then thought must be given to purchasing an expandable networked system.

7.7.5. Data Protection and the Law

Data protection legislation has become an important international issue. Such legislation is designed to prevent the abuse of personal data and to satisfy the requirements of international conventions such as the Council of Europe Convention. Both these points have become important because of the widespread use of databases and the requirements for importing and exporting data. If the Council of Europe Convention is not ratified by a country then the transfer of data to and from the non-complying country will become difficult as complying countries will refuse to deal with its computer bureaux. Typical conditions are that all personal data should be registered and none of these data should be used, disclosed or exported except as registered. Data users and computer bureaux must register and the data subject is permitted access to their records and may correct or erase the records if applicable.

The following principles are typical.

- Data must be obtained and processed fairly and lawfully.

- Data shall only be held and disclosed for lawful purposes.

- Data can only be used in a manner compatible with the stated purposes.

- Data must be relevant, adequate (not excessive) for the stated purpose.

- Data must be accurate and up-to-date.

- Information must only be held for as long as it is necessary.

- An individual must be informed if information is stored and have access without undue cost or delay (Deletion or Correction may follow)

- Appropriate security measures must be taken.

These principles can only be upheld if information is stored on data users. Such registration may include the following details

- Name and description of data user.

- Description of purpose and type of data.

- Source of data.

- Recipient of data.

- Countries to which the data may be exported.

- An address for the data subject to apply for information.

Various exemptions may exist depending on the country, particularly in the medical arena. Physical or mental health data may be exempt from subject access or the access may be via a censor such as the subject's doctor. If the data are used for research and the subject is not identified in the results then subject access may be denied. In medical emergencies the data may be disclosed despite protective legislation against disclosure. If the data are available by law or are stored in backup files then they may be totally exempt from additional legal restraints. In all cases the data users must check the legal situation as it pertains to them, as both national and international law changes and new precedents are set.

7.8. Practical Considerations

When a clinician designs, specifies or chooses a system the following points should be considered; user friendliness, reliability, conciseness, how well proven the system is, whether a significant improvement will be realised, cost effectiveness, fail-safe features and specialist computer staff requirements (Smith 1990).

The first five points determine whether or not the system will be used. The system must be user friendly otherwise most clinicians will not start using it. If the system is not reliable, or is too verbose then the clinician will not be willing to spend time on it. If the system does not have a proven track record most clinicians will be concerned about the learning period associated with a new product which may prove to be useless. Lastly there is no point in using any system if it introduces new problems without resolving any old problems.

Our final three points cover economics and safety. Cost effectiveness must be taken into account, including capital cost and running costs. In addition attention must be paid to the economic assumptions of decisions made by a computer system. For instance MRI scans for all out-patients would detect problems very early on but prove to be impractical economically. The system should also fail-safe so, for example, an error in 1% of cases will cost money not lives. Finally, costs involving employing computer specialists and external consultants or running training courses and tutorials should be built into an assessment of a system.

8

Hospital Safety

8.1. Electrical Safety

If we receive an electric shock there are basically two effects which occur. Firstly our nervous system may be excited, and secondly, we may suffer severe burns due to the resistive heating effect of the passage of current through our bodies. The stimulation of our nervous system may cause us injury through excitation of our muscles. However, as the heart is essentially a muscle, its stimulation represents the greatest risk through electrocution.

8.1.1. Levels of Electric Shock

Some individuals can sense currents as low as 100 µA externally applied at 50 Hz. However, other subjects may not be able to sense currents less than 0.5 mA. The threshold of feeling level for DC currents varies between 2 and 10 mA. These values tell us two things, that the threshold of feeling is an individual characteristic and that the body is more sensitive to AC signals than to DC signals. There is also a difference in the sensitivity of men and women to electric currents; women are generally more susceptible.

After the tingle, or threshold of feeling, the next level of electric shock is the 'let go' threshold. If you grasp a live conductor the muscles in your arm and hand are excited and contract. This causes you to grip the conductor more tightly whilst being electrocuted. At relatively low current levels you are able to overcome the current and still have voluntary control of your muscles. The approximate level above which most males cannot let go of a conducting object is 10 mA at 50 Hz. The level for women at this frequency is approximately 6 mA. At current levels above this limit there is severe pain and ligament damage may ensue. However, this level of electrocution is not life threatening unless the sufferer is in a hazardous situation. Perhaps a man is electrocuted on a ladder. In this instance a sudden muscle contraction may cause him to fall some distance. At a higher current still, the muscle contraction may be so violent as to cause fractures.

If the level of current flowing lies between 18 and 20 mA then there is potential for chest paralysis. If someone is electrocuted between points on the right hand and the right elbow, obviously this will not occur. If, however, a current passes across the chest of the patient then the muscles which control breathing may become frozen in the contracted state and therefore unable to function. Chest paralysis is extremely painful and sufferers soon become fatigued as they are unable to maintain an adequate supply of air.

If the current which passes across a subject's chest is greater than 22 mA but less than 75 mA, the normal beating rhythm of the heart may be disrupted. At currents greater than 75 mA but less than 400 mA, ventricular fibrillation may result. This occurs when the normal co-ordinated beating of the heart becomes disturbed and the heart quivers or shakes and no functional beating takes place. With currents greater than this level, the heart suffers sustained contraction, i.e. both ventricles and atria may contract and remain contracted. This is strangely less dangerous than ventricular fibrillation as, following removal of the stimulation, the heart starts beating in a co-ordinated fashion as the whole heart is simultaneously returned to its normal state. In ventricular fibrillation, each section of the heart beats in an uncoordinated fashion and the heart therefore requires an external stimulus to regain its co-ordination.

If a current greater than 10 A passes through the patient then, irrespective of nervous system damage, there will be serious burns due to the heating effect of the current. Accidents of this nature usually occur where high power cables or lines are used for industrial purposes.

8.1.2. Physical Differences in Electrocution

As we have already stated, men and women tend to have different thresholds for the physiological effects of electrocution. However, individuals within the same sex also have different thresholds. These depend to some extent on body weight and build. The skin resistance of a patient varies significantly with sweating and, therefore, a patient or a subject who touches a live cable with moist hands will receive a greater current than a patient with perfectly dry hands. However, while you are being electrocuted, you sweat. This itself reduces the skin's resistance and tends to increase the current flowing.

The path of the shock current through the sufferer's body determines the muscle groups and nerves affected. Obviously if a patient is electrocuted between two points on one side of the body, such that the current does not flow across their chest, the likelihood of serious injury to their heart or chest is reduced; whereas a current flow across the chest is potentially the most dangerous.

The duration of the current flow through the sufferer is related to the level of damage inflicted by the equation

$$I_{min} = \frac{116}{\sqrt{t}} \qquad (1)$$

Equation 1 is empirical and expresses the length of time or the minimum current which cause ventricular fibrillation. The minimum current for ventricular fibrillation is inversely proportional to the square root of time.

Figure 8.1 shows a graph of the threshold of feeling against frequency, it represents the frequency response of nervous tissue to alternating current. The effect of the current on the nervous tissue diminishes at frequencies below 10 Hz and at frequencies above 200 Hz. Unfortunately, the effect is maximum at approximately 50 Hz; in Europe and America the frequency chosen for domestic electricity supply is between 50 and 60 Hz, which correlates with the worst frequency for electric shock.

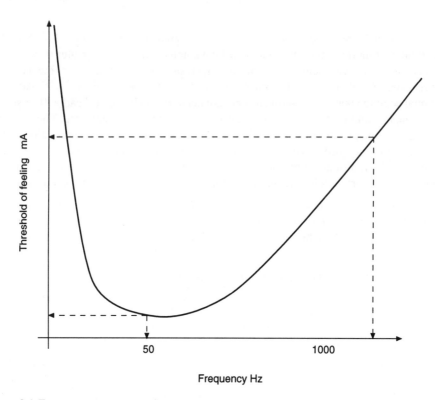

Figure 8.1 Frequency response of nervous tissue

8.1.3. Types of Electric Shock

An external electric shock is termed a macro shock. Potential danger to cardiac function exists with currents of greater than 10 mA. There are various situations which may result in a patient receiving a macro shock. Figure 8.3 shows two possible conditions in which a patient may receive a macro shock. In Figure 8.3a the live wire comes directly into contact with a patient connected lead. In Figure 8.3b the combination of a break in the ground connection and a live wire coming loose and touching the casing would cause macro shock if the patient touched the case.

An electric shock applied directly to the heart is termed a micro shock. Micro shocks almost exclusively occur in the clinical environment and involve equipment which is directly connected to the heart. For instance, if a patient's blood pressure is being monitored with a catheter transducer located in the heart, then there is the potential for micro shock. The threshold current of life threatening danger to cardiac function is as low as 50 μA for micro shock.

Increasing levels of micro shock cause various levels of disruption. The first level occurs when the natural rhythm of the heart becomes disturbed. Following this, there is pump failure, when the heart no longer supplies the blood flow required for the patient, thereafter ventricular fibrillation occurs. Obviously, patient data which determines these effects is sparse. However, a certain amount of this work has been conducted during heart operations. When a surgeon operates on a patient's heart, the heart is given a measured electric shock to elicit ventricular fibrillation, and facilitate work. Researchers have therefore been able to identify current thresholds relating to rhythm disturbance and ventricular fibrillation. The work has also been backed up by significant animal experimentation. A rhythm disturbance can be caused by a current as low as 80 μA, while 600 μA may cause ventricular fibrillation. The right atrium, where the sino-atrial node is situated, is the most susceptible part of the heart to electric shock.

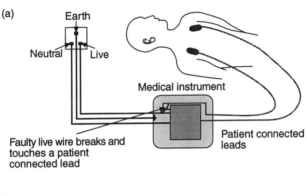

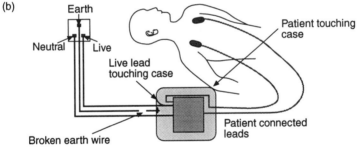

Figure 8.2 Macro shock possibilities

The threshold for feeling electric shock depends on the individual and circumstances, but lies between 100 and 500 μA. A danger of causing micro shock exists below the level of perception of the medical staff carrying out an investigation. Capacitive coupling from the live parts of an instrument can cause a current to flow to ground. This current is referred to as leakage current. Typical electronic instruments designed for industrial use may have leakage currents which although unnoticed by their users are above the levels which cause ventricular fibrillation if applied directly to the heart. Leakage currents are a major source of micro shock. The levels of leakage current which are permissible in medical equipment are strictly controlled. Any equipment brought in to a hospital from outside therefore represents a micro shock risk. In Figure 8.3a a patient with a cardiac catheter reaches to touch a mains powered radio. The radio designed for non medical use has a potentially dangerous leakage current which flows through the patient and to ground through the cardiac catheter.

The mains distribution system in hospitals uses three wires. The AC power is applied to two conductors, the live and neutral, whilst the third is connected to ground. The ground wire is commonly connected to metal screens within the instrument or to its case. If a live wire becomes loose and contacts such a screen the earth wire serves to carry fault current safely to ground and causes the circuit to fuse. The earth wire also carries leakage current due to capacitive coupling between the earthed screens in the instrument and its live parts. In the event of the earth conductor either in the power cord or the distribution system becoming broken this leakage current can no longer flow. A patient touching the faulty instrument case, as in Figure 8.3a, would therefore provide a path to ground for this current. In most cases this

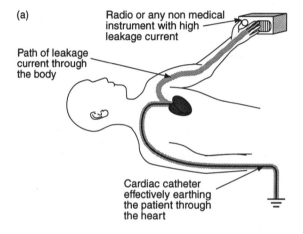

(a)

Radio or any non medical instrument with high leakage current

Path of leakage current through the body

Cardiac catheter effectively earthing the patient through the heart

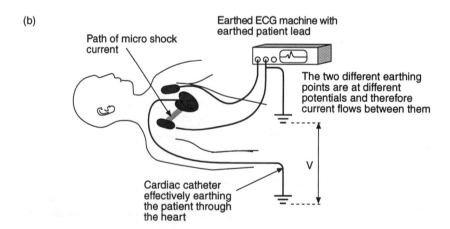

(b)

Path of micro shock current

Earthed ECG machine with earthed patient lead

The two different earthing points are at different potentials and therefore current flows between them

Cardiac catheter effectively earthing the patient through the heart

V

Figure 8.3 Micro shock situations

would not be noticed; however, if the patient has a cardiac catheter then the path to ground may be through the patient's heart. In this instance micro shock results.

The distribution system in older hospitals may have evolved rather than been designed. It is possible that separate power sockets in one room are connected to different earth points. These earth points may have different potentials. A patient connected to equipment with different earths receives a current flow owing to the potential difference. This is depicted in Figure 8.3b where the patient is simultaneously undergoing ECG measurement and heart catheterisation.

To minimise the risk of micro shock the majority of medical equipment used in intensive care areas incorporates isolated circuits, isolated power supplies, and earth free patient connections.

Patients during operations and in intensive care may require high concentrations of oxygen and other potentially explosive gases. These gases may build up in a fault condition and can be ignited by sparks from electrical equipment.

8.1.4. Isolated Power Supplies

The low thresholds for micro shock make the design of electrical equipment for clinical use difficult. Power supplies produced for industrial equipment have leakage currents in excess of those permitted for medical equipment. To construct equipment with leakage current levels acceptable in the medical environment it is necessary to use isolated power supplies whose primary and secondary windings are separated by an earthed screen (see Figure 8.4). The

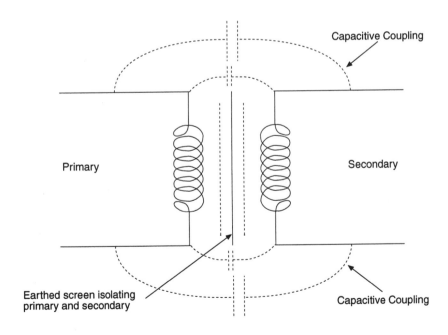

Figure 8.4 Isolated power supply

equipment must be constructed such that the capacitive coupling between the primary power supply parts and the secondary circuit and its connections is minimised. With careful design the leakage current from such a power supply can be reduced to below 25 μA. However, stray capacitance and hence leakage current can not be entirely eliminated.

8.1.5. Isolation Amplifiers

Isolation amplifiers allow two sections of a circuit at different potentials to be connected with a minimised leakage current flowing between them. In medical applications isolation amplifiers are used to protect the patient from both leakage currents and currents arising from fault conditions. They normally consist of a high impedance input section which must be followed by a low leakage barrier. This in turn is followed by a low impedance output, represented in Figure 8.5. There are three methods of transferring information from the input to the output via a low leakage barrier. They are transformer coupling, optical coupling and capacitive coupling. In medical applications capacitive barrier isolation amplifiers are not generally used.

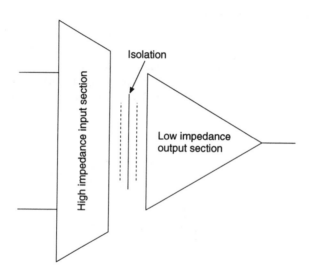

Figure 8.5 Isolation amplifiers

8.1.5.1. Transformer Isolation

The differential signal (see Figure 8.6) applied to the input of the amplifier is modulated and transmitted through the transformer. In the output section the signal is demodulated and amplified. Isolators fabricated in this way often incorporate feedback of the modulated signal to the input section via a second transformer winding to help correct for non-linear performance. A variety of modulation schemes is used including amplitude modulation.

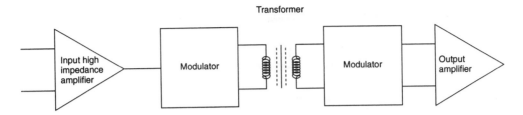

Figure 8.6 Transformer isolation

8.1.5.2. Optical Isolation

The barrier section is constructed using an optical source and detector. A Light Emitting Diode (LED) is used to transmit light to a photo diode used as a detector. The non-linear output characteristics and poor temperature stability of LEDs cause problems in the design of these devices. Similar modulation techniques to those used for transformer coupled amplifiers are employed (see Figure 8.7).

Both transformer and optically coupled isolation amplifiers incorporate a transformer winding to transmit power to the input section of the barrier. The device may also provide isolated power for pre-amplification stages. Isolated DC to DC converters are also used to supply isolated power to primary transducer circuits.

Isolation amplifiers must provide isolation up to approximately 5 kV before breakdown. The input stage of a isolated amplifier used in a bioelectronic recording system typically has a common mode rejection ratio of 120 dB.

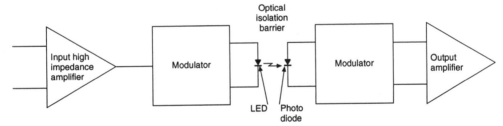

Figure 8.7 Optical isolation

8.1.6. Residual Current circuit breakers

In normal operation the current flowing down the live wire is equal to the return current flowing down the neutral wire. Discrepancies may be due to leakage currents. In fault conditions the current from the live wire may flow to ground through an alternative route. The current in the neutral conductor is then significantly less. Residual current circuit breakers sense the difference between the current flowing through the live and neutral wires of the supply and interrupt the supply if it exceeds a pre-determined limit.

Practical residual current detectors are built with a small symmetrical transformer placed in the live and neutral lines as shown in Figure 8.8. The live and neutral are wound in opposing directions. The flux produced by the respective coils cancels when the currents balance, so a sense coil measures no induced voltage. However, if the current in the neutral wire is different

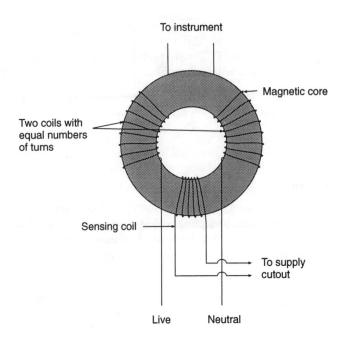

Figure 8.8 Residual current circuit breaker

from that flowing along the live wire then a net flux is induced and a voltage is produced in the sensing coil. If this induced voltage exceeds a pre-set limit a relay is switched to disable the supply. Residual current circuit breakers may be set to sense current differences of approximately 2 mA to protect against macro shock.

8.1.7. Electrical Safety Standards

Europe, Britain and America have standards for the electrical safety of medical equipment which specify acceptable levels of leakage current for a variety of grades of medical equipment. Equipment which is to be used in conjunction with instruments which are directly connected to the heart or equipment which makes a low impedance connection to the patient is classified as requiring higher levels of patient safety then equipment with high functional resistance at the point of application to the patient. It is important in all circumstances to design commercial and experimental research medical equipment to the safety standards in force in the intended country of use. Adherence to these standards ensures a minimum standard of safe operation is maintained. Failure to design to the safety standards may be evidence of negligence if the equipment should at any stage prove faulty or dangerous.

8.2. Radiation hazards

Dangers from radiation come from a number of sources. Around 20% of our normal dose of radiation comes from previously absorbed radioactive materials. As some materials localise in particular organs in the body these may be especially dangerous. Most of the remaining dose normally comes from background radiation, although in developed countries a significant proportion of this, when averaged through the population, comes from medical sources.

Additionally there is a risk of serious exposure of populations from radioactive accidents, although the authors of a UN report felt unable to make a rationally based assessment of the overall risk owing to a lack of data (UNSC, 1982). Biological damage is classified into two categories (in the same United Nations report):

1. Somatic effects, which apply directly to the irradiated individual, and cause tissue damage. The hazard from this source depends on the affected region of the body and the age of the individual (younger people are at greater risk owing to their higher rate of cell renewal).

2. Genetic effects which cause either gene mutation or chromosomal aberrations. The former are heritable alterations of the genetic material, which may either be dominant mutations causing effects on the immediate next generation, or recessive mutations, which may not express themselves for several generations to come.

Chromosomal aberrations result in a severely disrupted chromosomal make-up which may lead to very severe abnormalities.

A more accessible description (than is contained in the United Nations report) of biological effects of radiation, precautions, and legal requirements for radiation protection is given in *'An Introduction to Radiation Protection'* by A.Martin and A.B.Harbison (Chapman and Hall, 1979).

The somatic effect of radiation causes different forms of damage according to the absorbed dose. A dose of about 3 Gray causes death in 50% of individuals within 30 days of exposure. Death at this dose level is due to depletion of white blood cells resulting in a reduced resistance to infection. When proper medical attention can be given this cause of death may be significantly reduced. At significantly higher doses the survival time reduces to typically around 3–5 days. Death is due to serious loss of cells in the lining of the intestine, which is in turn followed by severe bacterial invasion. This is called *gastrointestinal death*.

In both these cases, the damage caused is roughly proportional to the dose absorbed. At lower level, the damage is termed *stochastic* since a probability of radiation induced damage can only realistically be calculated for a population. The primary form is carcinoma inducing, where signs of damage may become apparent many years after the exposure. A dose of 1 mSv given to each of a population of 1 million people gives rise to around 13 fatal cancers. The normal incidence of cancers in a population of this size per year is around 2000.

8.2.1. Basic precautions

The major consideration when dealing with ionising radiation must always be to consider whether the risks involved in its use may outweigh any possible benefits. There must be a clear strategy to minimise any exposures to radiation: this may be by using shielding, keeping a good distance and minimising the duration of any exposure. In the case of medical exposures particularly, it may be possible to reduce the exposure of organs not under investigation by using addition shielding and restricting the width of the beam. Sensitive areas of the body should be particularly avoided, such as the gonads, as should exposure of children and pregnant women.

When a source is used with a restricted beam, it should be ensured that the radiation which escapes the working area is minimised by shielding and distance: the subsequent movement of the source may require to be restricted so that it may not be used inadvertently in an unprotected direction. Particular care should be taken of workers in radiology as they are likely to be exposed to radiation much more frequently than are their patients.

8.2.2. Legal requirements

Regulations for the use of radiation are not the same in all countries: there is increasing uniformity in the European Community as certain of the Euratom codes of practice are adopted. The major legislation in the UK is the Health and Safety at Work Act (1974) which defines responsibilities for safety. The statutory body which supervises the enactment of this legislation, the Health and Safety Executive, has drawn up appropriate regulations regarding the designation of types of facility in relation to their risk of causing exposure.

The UK National Radiological Protection Board is responsible for the acquisition of knowledge in the field of radiation safety and providing services to assist in that end.

Both of these topics are covered in greater detail in Martin and Harbison: anyone requiring to use ionising radiation should however become fully conversant with their appropriate legal codes.

8.3. Ultrasound safety

Although ultrasound is widely regarded as being harmless, it is possible to weld materials using ultrasound and to destroy kidney stones in situ. It would be more accurate to say that the risk to patients of diagnostic ultrasound at the intensities currently used is minimal.

The passage of ultrasound through tissue causes heating as its wave energy is converted to thermal energy through relaxation processes. The heating effect of ultrasound is used for some therapeutic applications. In Doppler or echo imaging systems this process is unwanted. In sensitive tissues, such as the brain, the heating effect of ultrasound could be dangerous. However, the intensities used in current diagnostic equipment are such that the heating effect is negligible.

As ultrasound travels through tissue, the density compaction and rarefaction caused may reach energy levels after which cavitation occurs. It most probably would occur in continuous fields of high intensity possibly caused by standing waves. A number of mechanisms which are potentially damaging have been identified, and studies have been undertaken into the possible systemic effects of damage due to diagnostic ultrasound. They are not believed to occur at energy levels below 100 mW cm^{-2} (see Evans, 1989).

References

ALLEN, L. FRIEDER, O. 1992: Exploiting Database Technology in the Medical Arena: A critical Assessment of integrated systems for picture archiving and communications. *IEEE Engineering in Medicine and Biology* 11, 42-49.

ANNIS, R.J. HOLTON, J.W. 1989: Balancing the need to expand financial service while reducing staffing levels by utilising End-user micro computer services. *Proceedings of the 22nd Hawaii International Conference on System Sciences Emerging Technologies and Applications Track* 4, 61-69.

ARMSTRONG, R.F. BULLEN, C. COHEN, S.L. SINGER, M. WEBB, A.R. 1992: *Critical Care Algorithms* Oxford University Press.

BARR, A. FEIGENBAUM, E.A. 1986: *The Handbook of Artificial Intelligence* Addison Wesley Vol 2 177-222.

BARRIE, J.L. MARSH, D.R. 1992: Quality of data in the Manchester orthopaedic database. *British Medical Journal* 304 159-162.

BERNAL, J.D. 1957: *Science in History*, Watts.

BLACK, U.D. 1989: *Data Networks*. Prentice-Hall.

BLEANEY, B., BLEANEY, B.I. 1976: *Electricity and Magnetism,* Oxford .

BRACEWELL, R.N. 1986: *The Fourier Transform and its Applications*, 2nd Edition, McGraw-Hill

BRODY, W.R. 1983: *Scanned Projection Radiography: physical and clinical aspects*: Digital Radiology Clinical and Physical Aspects, IPSM London

CARMICHAEL, J.H.E, 1988: *Protection of the Patient in Diagnostic Radiology - ICRP Philosophy*, IPSM Report 55.

CAROLA, R., HARTLEY, J.P. AND NOBACK, C.R. 1990: *Human Anatomy and Physiology*, 3rd Edition, McGraw-Hill

DENNING, D.E.R. 1982: *Cryptography and Data Security* Addison-Wesley.

DUCK, F.A. 1990: *Physical Properties of Tissue: a comprehensive reference book*, Academic Press

ELLIS, B.W. MICHIE, H.R. ESUFALI, S.T. PYPER, R.J.D. DUDLEY, H.A.F. 1987: Development of a microcomputer-based system for surgical audit and patient administration : a review. *Journal of the Royal Society of Medicine,* 80.

EVANS, D.H. MCDICKEN, W.W.SKIDMORE, R. WOODCOCK, J.P. 1989: *Doppler Ultrasound,* Wiley

FEISCHI, M, 1984: *Artificial Intelligence in Medicine: Expert Systems* Chapman and Hall Translated by Cramp, D. 1990.

FERREIRA, D.P. WILSON, J.R. 1990: IMPLANTOR- An Intelligent Tutoring System for Orthopaedic Repair. *IEE Colloquium on Intelligent Decision Support Systems & Medicine Digest,* 143 11/1-4 .

FUMAI, N. COLLET, C. PETRONI, M. ROGER, K. LAM, A. SAAB, E. MALOWANY, A.S. CARNEVALE, F.A. GOTTESMAN, R.D. 1991: Database Design of an Intensive Care Unit Patient Data Management System. *Computer Based Medical Systems Proceedings of the fourth annual IEEE symposium 78-85*

GONZALEZ, R.C. and WOODS, R.E. 1992: *Digital Image Processing,* Addison-Wesley.

GREENING, J.R. 1981: *Fundamentals of Radiation Dosimetry,* Adam Hilger.

HENKIND, S.J. HARRISON, M.C. 1988: An Analysis of Four Uncertainty Calculi. *IEEE Transactions on Systems, Man and Cybernetics* **18** 5.

HILL, C.R. (ed.) 1986: *Physical Principles of Medical Ultrasound,* Ellis Horwood.

HOY, D. 1990: Computer-Assisted Care Planning. *Nursing Times* **86** 9.

KING, K. 1990: A Model Based Toolkit for Building Medical Diagnostic Support Systems in Developing Countries. *IEE Colloquium on Intelligent Decision Support Systems & Medicine Digest,* 143 9/1-16

KRIEGER, D. BURK, G. SCLABASSI, R.J. 1991: Neuronet: A Distributed Real-Time System for monitoring Neurophysiologic Function in the Medical Environment. *Computer* **24** 3 45-55.

LANE, V. P.1985: *Security of computer based information systems.* Macmillan.

LATHI, B.P 1983: *Modern Digital and Analogue Communication Systems,* Holt Reinhardt and Winston.

LERSKI, R.A. 1985: *Physical Principles and Clinical Applications of Nuclear Magnetic Resonance,* IPSM.

LONGHURST, R.S. 1967: *Geometrical and Physical Optics,* Longmans.

MACOVSKI, A. 1983: *Medical Imaging Systems,* Prentice-Hall.

MCCOLLUM, P.T. SUSHIL, K.G. MANTESE, V.A. JOSEPH, M. KARPLUS, E. GRAYWEALE, A.C. SHANIK, G.D. LIPPEY, E.R. DEBURGH, M.M. LUSBY, R.J. 1990: Microcomputer Database and System of Audit for the Vascular Surgeon. *Aust N.Z. J. Surg,* 60 519-523

OLAGUNJU, D.A. GOLDENBERG, I.F. 1989: Clinical Data Bases: Who Needs One (Criteria Analysis). *Proceedings of the Second Annual IEEE Symposium on Computer-Based Medical Systems,* 36-39

RAMONI, M. STEFANELLI, M. MAGNANI, L. BARSOI, G. 1992: An Epistemological Framework for Medical Knowledge -Based Systems. *IEEE Transactions on Systems, Man and Cybernetics* 22 1361-1375.

RICH, E. 1991: *Artificial Intelligence,* McGraw-Hill.

RODNICK, J.E. 1990: An opposing View. *The Journal of Family Practice,* 30 460-464 .

SHORTCLIFFE, E.H. 1976: *Computer-Based Medical Consultations: MYCIN.* Artificial Intelligence Series, Elsevier .

SMITH, M. 1990: Intelligent Health Care Information Systems: Are they Appropriate? *IEE Colloquium on Intelligent Decision Support Systems & Medicine Digest* No 143 pp 3/1-3.

SPANN, S.J. 1990: Should the Complete Medical Record Be Computerised in Family Practice? An Affirmative View. *The Journal of Family Practice* 30 457-460.

SZOLOVITS, P. 1982: Artificial Intelligence in Medicine. *AAAS Selected Symposium* **51**, West View Press.

TANNENBAUM, A.S. 1989: *Computer Networks,* Prentice-Hall.

UNSC, 1982: United Nations Scientific Committee on the Sources and Biological Effects of Ionising Radiation.

WEBB, S.(ed) 1990: *The Physics of Medical Imaging*, Adam Hilger.

WELLS, P.N.T. 1977: *Biomedical Ultrasonics*, Academic Press

WILLIAMS, B.T. 1982: *Computer Aids to Clinical Decisions*, Vol I Chap 2 CRC Press.

Index